BURP SUITE

NOVICE TO NINJA

PEN TESTING CLOUD, NETWORK, MOBILE & WEB APPLICATIONS

4 BOOKS IN 1

BOOK 1
BURP SUITE FUNDAMENTALS: A NOVICE'S GUIDE TO WEB APPLICATION SECURITY

BOOK 2
MASTERING BURP SUITE: PEN TESTING TECHNIQUES FOR WEB APPLICATIONS

BOOK 3
PENETRATION TESTING BEYOND WEB: NETWORK, MOBILE & CLOUD WITH BURP SUITE

BOOK 4
BURP SUITE NINJA: ADVANCED STRATEGIES FOR ETHICAL HACKING AND SECURITY AUDITING

ROB BOTWRIGHT

Copyright © 2023 by Rob Botwright
All rights reserved. No part of this book may be reproduced or transmitted in any form or by any means, electronic or mechanical, including photocopying, recording, or by any information storage and retrieval system, without permission in writing from the publisher.

Published by Rob Botwright
Library of Congress Cataloging-in-Publication Data
ISBN 978-1-83938-567-4
Cover design by Rizzo

Disclaimer

The contents of this book are based on extensive research and the best available historical sources. However, the author and publisher make no claims, promises, or guarantees about the accuracy, completeness, or adequacy of the information contained herein. The information in this book is provided on an "as is" basis, and the author and publisher disclaim any and all liability for any errors, omissions, or inaccuracies in the information or for any actions taken in reliance on such information. The opinions and views expressed in this book are those of the author and do not necessarily reflect the official policy or position of any organization or individual mentioned in this book. Any reference to specific people, places, or events is intended only to provide historical context and is not intended to defame or malign any group, individual, or entity. The information in this book is intended for educational and entertainment purposes only. It is not intended to be a substitute for professional advice or judgment. Readers are encouraged to conduct their own research and to seek professional advice where appropriate. Every effort has been made to obtain necessary permissions and acknowledgments for all images and other copyrighted material used in this book. Any errors or omissions in this regard are unintentional, and the author and publisher will correct them in future editions.

TABLE OF CONTENTS – BOOK 1 - BURP SUITE FUNDAMENTALS: A NOVICE'S GUIDE TO WEB APPLICATION SECURITY

Introduction 5
Chapter 1: Introduction to Web Application Security 9
Chapter 2: Getting Started with Burp Suite 17
Chapter 3: Configuring Burp Suite for Your Needs 27
Chapter 4: Understanding HTTP and Web Protocols 34
Chapter 5: Scanning and Crawling with Burp Suite 41
Chapter 6: Intercepting and Modifying Web Requests 48
Chapter 7: Analyzing and Exploiting Vulnerabilities 54
Chapter 8: Web Application Authentication Testing 62
Chapter 9: Advanced Burp Suite Techniques 69
Chapter 10: Reporting and Remediation 77

TABLE OF CONTENTS – BOOK 2 - MASTERING BURP SUITE: PEN TESTING TECHNIQUES FOR WEB APPLICATIONS

Chapter 1: Burp Suite Essentials and Setup 88
Chapter 2: Web Application Reconnaissance 97
Chapter 3: Identifying and Exploiting Web Vulnerabilities 105
Chapter 4: Advanced Scanning and Crawling Techniques 114
Chapter 5: Intercepting and Analyzing Web Traffic 123
Chapter 6: Web Application Authentication and Authorization 130
Chapter 7: Attacking Client-Side Security 138
Chapter 8: Exploiting API and Web Services 144
Chapter 9: Advanced Burp Suite Automation 152
Chapter 10: Reporting and Post-Exploitation 160

TABLE OF CONTENTS – BOOK 3 - PENETRATION TESTING BEYOND WEB: NETWORK, MOBILE & CLOUD WITH BURP SUITE

Chapter 1: Expanding Your Penetration Testing Horizons 170
Chapter 2: Setting Up Burp Suite for Diverse Targets 178
Chapter 3: Network Penetration Testing with Burp Suite 185
Chapter 4: Mobile Application Assessment with Burp Suite 191
Chapter 5: Securing Cloud Environments with Burp Suite 198
Chapter 6: Advanced Reconnaissance Techniques 205
Chapter 7: Exploiting Network Vulnerabilities 213
Chapter 8: Mobile App Exploitation and Reverse Engineering 220
Chapter 9: Cloud Security Assessment and Hardening 224
Chapter 10: Automating Multi-Platform Assessments 232

TABLE OF CONTENTS – BOOK 4 - BURP SUITE NINJA: ADVANCED STRATEGIES FOR ETHICAL HACKING AND SECURITY AUDITING

Chapter 1: The Path to Burp Suite Mastery 237
Chapter 2: Advanced Burp Suite Configuration and Customization 244
Chapter 3: Leveraging Burp Macros and Extensions 249
Chapter 4: Mastering Burp's Intricate Scanning Techniques 253
Chapter 5: Exploiting Advanced Web Application Vulnerabilities 260
Chapter 6: Client-Side Attacks and Beyond 264
Chapter 7: Network and Infrastructure Hacking with Burp Suite 272
Chapter 8: Beyond Web: Cloud, Mobile, and IoT Security 277
Chapter 9: Burp Suite in Enterprise Environments 285
Chapter 10: Reporting, Remediation, and Staying Ahead 289
Conclusion 298

Introduction

Welcome to the ultimate journey of becoming a cybersecurity expert, a master of ethical hacking, and a guardian of digital fortresses. In this immersive book bundle, "Burp Suite: Novice to Ninja - Pen Testing Cloud, Network, Mobile & Web Applications," we will embark on an extraordinary adventure through the ever-evolving landscape of cybersecurity.

In today's interconnected world, the importance of securing digital assets cannot be overstated. Whether you're safeguarding a web application, fortifying network defenses, assessing the security of mobile devices, or ensuring the integrity of cloud environments, you'll find the knowledge and skills you need within these pages.

This bundle consists of four distinct volumes, each designed to take you from a novice explorer to a seasoned ninja in the realm of ethical hacking and penetration testing. Let's take a closer look at what awaits you in each book:

Book 1 - Burp Suite Fundamentals: A Novice's Guide to Web Application Security: We begin our journey at the foundation of web application security. This book is your trusty map as you navigate the intricate world of web vulnerabilities. From understanding the basics to harnessing the power of Burp Suite, you'll gain the insights needed to uncover and mitigate threats effectively.

Book 2 - Mastering Burp Suite: Pen Testing Techniques for Web Applications: Building on the knowledge acquired in the first book, we dive deeper into the art of ethical hacking. Armed with advanced techniques and insider tips, you'll become proficient in leveraging Burp Suite to identify vulnerabilities, execute precise attacks, and secure web applications against potential threats.

Book 3 - Penetration Testing Beyond Web: Network, Mobile & Cloud with Burp Suite: Our journey extends beyond web applications as we venture into the domains of network, mobile, and cloud security. Discover how Burp Suite can be adapted to address a broader spectrum of challenges, equipping you to assess and fortify various digital landscapes.

Book 4 - Burp Suite Ninja: Advanced Strategies for Ethical Hacking and Security Auditing: In the final leg of our expedition, we ascend to the status of security auditors and ethical hacking ninjas. Armed with advanced strategies, customization techniques, scripting, and automation, you'll not only identify vulnerabilities but also craft comprehensive security reports and devise effective remediation strategies.

Throughout this bundle, you'll find a friendly guide accompanying you on this exhilarating journey. With each turn of the page, you'll gain new insights, practical *skills, and the confidence to tackle cybersecurity challenges head-on. Whether you're an aspiring cybersecurity professional, a seasoned expert seeking to expand your knowledge, or anyone in between, this bundle has something valuable to offer.*

So, prepare to don your virtual armor, sharpen your digital sword, and embark on this epic quest toward becoming a cybersecurity champion. The world of ethical hacking and security auditing awaits your arrival, and with "Burp Suite: Novice to Ninja," you'll be well-prepared to navigate its intricate paths and conquer its formidable challenges.

BOOK 1
BURP SUITE FUNDAMENTALS
A NOVICE'S GUIDE TO WEB APPLICATION SECURITY

ROB BOTWRIGHT

Chapter 1: Introduction to Web Application Security

Web application security is a critical aspect of cybersecurity in the digital age. It plays a pivotal role in safeguarding sensitive data, protecting user privacy, and ensuring the integrity and availability of online services. In today's interconnected world, where businesses and individuals rely heavily on web applications for various purposes, the importance of web application security cannot be overstated.

The ubiquity of web applications has made them a prime target for cybercriminals. These malicious actors often seek to exploit vulnerabilities within web applications to gain unauthorized access, steal sensitive information, or disrupt critical services. Therefore, understanding and prioritizing web application security is imperative for both organizations and individuals.

One of the key reasons why web application security is crucial is the vast amount of sensitive data that flows through these applications. From personal information such as names and addresses to financial data like credit card numbers, web applications handle a treasure trove of valuable data. Any breach or compromise of this data can have severe consequences, including financial loss, identity theft, and damage to an organization's reputation.

Web application security is not just about protecting data; it's also about ensuring the availability of online services. Downtime caused by attacks or vulnerabilities can lead to lost revenue, disrupt user experiences, and erode trust. Businesses that rely on web applications for e-commerce, communication, or customer engagement simply cannot afford extended periods of unavailability.

Furthermore, web application security is vital for maintaining the trust of users and customers. When individuals use a web application, they trust that their data will be handled responsibly and securely. A breach of this trust can result in users abandoning a service, potentially causing significant harm to a business.

To address these challenges and mitigate risks, organizations and security professionals employ various strategies and tools. One of the foundational tools in the arsenal of web application security is Burp Suite. This powerful software suite is designed to help identify, assess, and remediate vulnerabilities in web applications.

Burp Suite provides a comprehensive set of features that enable security professionals to analyze web traffic, intercept and modify requests, and identify security weaknesses. With its user-friendly interface and robust scanning capabilities, Burp Suite empowers security experts to uncover vulnerabilities such as SQL injection, cross-site scripting (XSS), and security misconfigurations.

Web application security encompasses a wide range of threats and attack vectors. One common threat is SQL injection, where an attacker manipulates input fields to inject malicious SQL queries into a web application's database. This can lead to unauthorized access and data leakage. Burp Suite aids in detecting and preventing SQL injection by analyzing and sanitizing input data.

Cross-site scripting (XSS) is another prevalent threat in web applications. It occurs when attackers inject malicious scripts into web pages viewed by other users. This can result in session hijacking, data theft, or other malicious activities. Burp Suite assists in identifying and remediating XSS vulnerabilities by scanning web application code and responses for suspicious script injections.

Security misconfigurations are another area of concern. They occur when web applications are not properly configured, leaving them vulnerable to attacks. Burp Suite helps security professionals identify misconfigurations by conducting comprehensive scans of web applications, flagging potential issues for remediation.

Beyond these specific threats, Burp Suite is a versatile tool for assessing overall web application security. It allows security experts to map out the entire application, identify all accessible pages and endpoints, and assess potential attack vectors. This holistic approach helps security professionals gain a comprehensive view of a web application's security posture.

Web application security is not a one-time endeavor but an ongoing process. Threats evolve, new vulnerabilities emerge, and web applications change over time. Therefore, continuous monitoring and testing are essential components of web application security. Burp Suite supports this aspect by providing automation and scripting capabilities, allowing security professionals to conduct regular scans and assessments.

In addition to its scanning and testing capabilities, Burp Suite offers features for manual testing and analysis. Security experts can intercept and manipulate web traffic, analyze request and response headers, and test various input fields for vulnerabilities. This hands-on approach enhances the effectiveness of security assessments and enables the discovery of complex vulnerabilities.

Moreover, Burp Suite is equipped with advanced features for handling complex scenarios. It can be integrated with other security tools and technologies, allowing for seamless workflows in enterprise environments. This level of flexibility and extensibility makes Burp Suite a valuable asset for security professionals operating in diverse and challenging settings.

While Burp Suite is a powerful tool, it is essential to emphasize that expertise is equally crucial in web application security. Security professionals must possess a deep understanding of web technologies, programming languages, and attack vectors. They must keep abreast of emerging threats and vulnerabilities and continuously refine their skills to stay ahead of cybercriminals.

Web application security is a dynamic field that requires constant vigilance and adaptation. Organizations that invest in robust security practices, including the use of tools like Burp Suite, can significantly reduce the risk of security breaches and protect their valuable assets. However, it is essential to remember that security is a shared responsibility, and everyone who uses web applications has a role to play in maintaining a secure online environment.

In summary, web application security is of paramount importance in today's digital landscape. It safeguards sensitive data, ensures the availability of online services, and maintains the trust of users and customers. Burp Suite is a valuable tool that aids security professionals in identifying and mitigating web application vulnerabilities. However, effective web application security requires a holistic approach, including continuous monitoring, testing, and ongoing education. By prioritizing web application security and leveraging tools like Burp Suite, organizations and individuals can navigate the ever-evolving threat landscape with confidence. Web application security is a critical concern in today's digital landscape, as web applications are integral to our daily lives and business operations. These applications, while providing numerous benefits, are also susceptible to a wide range of security threats that can have serious consequences if not addressed. Next, we will explore some of the most common web application security threats that organizations and individuals face. One of the most prevalent threats is SQL injection, a technique where attackers manipulate input fields to inject malicious SQL queries into a web application's database. This can result in unauthorized access to sensitive data, data leakage, or even the complete compromise of a web application. Cross-site scripting (XSS) is another widespread threat, where attackers inject malicious scripts into web pages that are then executed by other users' browsers. XSS can lead to session hijacking, data theft, and the defacement of web pages, eroding user trust. Authentication and session management vulnerabilities are also common, as attackers often target weak authentication mechanisms or exploit flaws in session management to gain unauthorized access. Insecure direct object references (IDOR) are a type of vulnerability where attackers can manipulate input to access other users' data or resources. This can lead to data exposure and privacy breaches. Security misconfigurations are another frequent issue, resulting from improper or incomplete configuration of web applications or their supporting infrastructure. Attackers can exploit these misconfigurations to gain unauthorized access or disrupt services. Cross-Site Request Forgery (CSRF) attacks are designed to trick users into performing unwanted actions in their authenticated sessions, potentially leading to unauthorized changes or actions. Web application firewalls (WAFs) are commonly used to protect against various web application threats. WAFs inspect incoming traffic and filter out malicious requests, offering a layer of defense against attacks like SQL injection and XSS. In addition to these threats, the OWASP (Open Web Application Security Project) Top Ten Project identifies and ranks the most critical web application security risks. This list provides valuable insights into the key challenges faced by organizations and security professionals. The OWASP Top Ten includes vulnerabilities such as injection attacks, broken authentication, sensitive data exposure, XML external entities (XXE), and more. Understanding these common threats is essential for anyone involved in web application development, testing, or security. Mitigating these threats requires a multi-faceted approach that combines secure coding practices, regular security assessments, and the use of security tools like Burp Suite. SQL injection, one of the most prevalent threats, can be mitigated by using prepared statements and parameterized queries to ensure that user input is properly sanitized before interacting with a database. Cross-site scripting (XSS) vulnerabilities can be prevented by validating and escaping user-generated content and employing security headers like Content Security Policy (CSP). Authentication and session management vulnerabilities can be addressed by implementing secure authentication mechanisms, enforcing strong password policies, and implementing proper session management controls. To protect against insecure direct object references (IDOR), applications should perform proper access control checks and avoid exposing internal references directly in URLs. Security misconfigurations can be mitigated through regular security assessments, automated scanning tools, and adherence to security best practices for server and application configurations. Cross-Site Request Forgery (CSRF) attacks can be prevented by using anti-CSRF tokens and ensuring that sensitive actions require user consent. In addition to these preventive measures, ongoing monitoring and threat detection are crucial. Web application firewalls (WAFs) can help detect and block attacks in real-time, offering an additional layer of defense. Regular security assessments and penetration *testing, often conducted with tools like Burp Suite, can uncover vulnerabilities and weaknesses that need attention. Moreover, organizations should stay informed about the latest security threats and vulnerabilities by participating in the security community, attending conferences, and following industry news and updates. Collaboration among developers, testers, and security professionals is essential for effectively addressing web application security threats. Developers should receive training on secure coding practices and be encouraged to incorporate security into their development processes from the outset. Security teams should work closely with development teams to identify and prioritize vulnerabilities and ensure that appropriate remediation measures are taken. Ultimately, web application security is an ongoing effort that requires diligence and a proactive approach. Organizations and individuals must remain vigilant, adapt to evolving threats, and continuously improve their security posture. By understanding and addressing common web application security threats, we can better protect sensitive data, maintain user trust, and ensure the secure operation of web applications in an increasingly interconnected world.*

Chapter 2: Getting Started with Burp Suite

Next, we will delve into the essential topic of installing and setting up Burp Suite, a powerful tool for web application security testing. Before we can begin using Burp Suite effectively, it's crucial to ensure that it's correctly installed and configured on your system. Burp Suite is available in both free and paid versions, with the free version offering many valuable features for security professionals and enthusiasts. To get started, visit the official PortSwigger website to download the appropriate version of Burp Suite for your operating system. Once the download is complete, follow the installation instructions provided on the website to install Burp Suite on your machine. The installation process typically involves running the installer package and selecting the installation directory. After installation, you can launch Burp Suite from your system's applications menu or by executing the appropriate command in the terminal. Upon starting Burp Suite, you will be greeted with a welcoming screen, and the application will begin initializing. Burp Suite is a Java-based application, so you must have Java Runtime Environment (JRE) installed on your system to run it. If you don't have JRE installed, you can download and install it from the official Oracle website or use an open-source alternative like OpenJDK. Once Burp Suite is up and running, it's time to configure it to suit your specific testing needs and preferences. Burp Suite offers a wide range of configuration options, allowing you to customize various aspects of the tool. To access the configuration settings, click on the "User options" button in the toolbar or navigate to the "Project options" tab. Within the configuration settings, you can define proxy options, configure target scope, set up your preferred browser, and adjust various other parameters. Proxy configuration is a critical aspect of Burp Suite setup, as it allows the tool to intercept and analyze web traffic between your browser and the target web application. Burp Suite acts as a proxy server, sitting between your browser and the web application, and capturing all HTTP requests and responses. To configure your browser to use Burp Suite as a proxy, you'll need to modify your browser's proxy settings. The proxy settings typically include specifying the host (localhost or the IP address where Burp Suite is running) and the port number (by default, Burp Suite uses port 8080). Once your browser is configured to use Burp Suite as a proxy, you can start intercepting and analyzing web traffic by enabling the interception feature in Burp Suite. Burp Suite's interception tool allows you to selectively intercept and modify HTTP requests and responses, giving you full control over the traffic between your browser and the web application. Before you start intercepting traffic, it's a good practice to define a target scope in Burp Suite. The target scope helps you narrow down your testing focus to specific domains, URLs, or web applications, ensuring that you only intercept and assess the traffic that is relevant to your testing objectives. To configure the target scope, navigate to the "Target" tab in Burp Suite and add the domains or URL patterns that you want to include or exclude from your testing scope. Another essential aspect of Burp Suite setup is configuring your preferred browser for testing. Burp Suite provides instructions for configuring various popular browsers, such as Firefox, Chrome, and Safari, to work seamlessly with the tool. These instructions typically involve installing browser extensions or configuring proxy settings within the browser itself. Once your browser is configured, you can use it to navigate to the web application you want to test while Burp Suite intercepts and analyzes the traffic in the background. Burp Suite also offers the option to use its built-in web browser for testing, which can be convenient for certain scenarios. The built-in browser is preconfigured to work with Burp Suite, eliminating the need for additional browser setup. However, using the built-in browser may not always replicate the behavior of real-world browsers, so it's essential to consider your testing requirements when choosing your testing environment. Now that Burp Suite is correctly installed, configured, and ready for action, it's time to explore its various features and capabilities. Burp Suite provides a user-friendly interface with a variety of tools and tabs designed to assist security professionals in every aspect of web application testing. The main components of the Burp Suite interface include the Target, Proxy, Spider, Scanner, Intruder, Repeater, Sequencer, Decoder, Comparer, Extender, and Project options tabs. Each of these tabs serves a specific purpose, allowing you to perform various tasks, from mapping the application to identifying vulnerabilities and automating attacks. The Target tab provides an overview of the target scope you defined earlier, allowing you to manage the list of included and excluded domains and URLs. It also provides information about the target's site map, which is a hierarchical representation of all the pages and resources Burp Suite has encountered during testing. The Proxy tab is where you can intercept and manipulate HTTP requests and responses between your browser and the target web application. This tab is essential for understanding how web applications work and for identifying vulnerabilities such as SQL injection and cross-site scripting (XSS). The Spider tab allows you to crawl the target web application to discover and map its content and functionality. Crawling is a crucial step in understanding the structure of the application and identifying potential entry points for testing. The Scanner tab is where Burp Suite's automated vulnerability scanner comes into play. This scanner can detect a wide range of web application vulnerabilities, including SQL injection, XSS, and security misconfigurations. The Intruder tab is a powerful tool for automating attacks on web applications. It allows you to define and execute customized attack scenarios, making it invaluable for testing the security of input fields and parameters. The Repeater tab lets you interact with individual HTTP requests and responses, allowing for manual testing and verification of vulnerabilities. It is particularly useful for fine-tuning and exploring potential weaknesses. The Sequencer tab assists in testing the randomness and unpredictability of *tokens and session identifiers used by web applications. By analyzing the quality of randomness, you can identify weaknesses that could be exploited by attackers. The Decoder tab provides various encoding and decoding functions, enabling you to manipulate data in different formats. This can be useful for understanding how web applications handle user input and for crafting payloads for attacks. The Comparer tab allows you to compare two pieces of data, which can be helpful for identifying subtle differences in responses or identifying vulnerabilities. The Extender tab is where you can extend Burp Suite's functionality by adding custom extensions and scripts. These extensions can enhance your testing capabilities and automate repetitive tasks. Finally, the Project options tab is where you can configure project-specific settings, including session handling rules, authentication details, and scan policies. By navigating through these tabs and exploring their respective features, you'll gain a deeper understanding of how to leverage Burp Suite for effective web application security testing. Burp Suite is a versatile tool that can be used for a wide range of web application security assessments, from manual testing and verification to automated scanning and exploitation. As you become more familiar with the tool and its capabilities, you'll be better equipped to identify and address web application vulnerabilities, ultimately improving the security posture of the web applications you assess. With the installation and setup of Burp Suite complete, you're now ready to embark on your journey into the world of web application security testing, armed*

with a powerful tool and the knowledge to make the most of it.
Next, we will explore the Burp Suite interface, a powerful and feature-rich environment designed to assist security professionals in web application security testing. As you begin your journey with Burp Suite, understanding how to navigate its interface is crucial to effectively utilize its wide range of tools and capabilities. The Burp Suite interface is user-friendly and well-organized, making it relatively easy to find and access the features you need. The main window is divided into several tabs, each dedicated to a specific aspect of web application testing. The primary tabs include Target, Proxy, Spider, Scanner, Intruder, Repeater, Sequencer, Decoder, Comparer, Extender, and Project options. These tabs provide access to various tools and functionalities designed to assist you in different stages of your security assessments. The Target tab is your starting point for managing the target scope of your assessment. Here, you can specify the domains and URLs you want to include or exclude from your testing scope. The Target tab also provides a site map, which is a hierarchical representation of all the pages and resources encountered during your assessment. The Proxy tab is where you can intercept and manipulate HTTP requests and responses between your browser and the target web application. It allows you to inspect, modify, and forward traffic, giving you complete control over the communication between your browser and the web application. The Spider tab is essential for discovering and mapping the content and functionality

of the target web application. It enables you to crawl the application, identify potential entry points, and gain insights into the application's structure. The Scanner tab is home to Burp Suite's automated vulnerability scanner, which can detect a wide range of web application vulnerabilities. These vulnerabilities include SQL injection, cross-site scripting (XSS), and security misconfigurations. The Intruder tab is a versatile tool for automating attacks on web applications. You can use it to define and execute customized attack scenarios on input fields and parameters, making it invaluable for testing security vulnerabilities. The Repeater tab provides a way to interact with individual HTTP requests and responses, allowing for manual testing, verification, and fine-tuning of potential weaknesses. It's a useful tool for exploring and validating vulnerabilities discovered during testing. The Sequencer tab assists in assessing the randomness and unpredictability of tokens and session identifiers used by web applications. By analyzing the quality of randomness, you can identify weaknesses that attackers might exploit. The Decoder tab offers various encoding and decoding functions, enabling you to manipulate data in different formats. This can be helpful for understanding how web applications handle user input and crafting payloads for attacks. The Comparer tab allows you to compare two pieces of data, which can be valuable for identifying subtle differences in responses or pinpointing vulnerabilities. The Extender tab is where you can extend Burp Suite's functionality by adding custom extensions and scripts. These extensions can enhance your testing capabilities and automate repetitive tasks, tailoring Burp Suite to your specific needs. Finally, the Project options tab allows you to configure project-specific settings, including session handling rules, authentication details, and scan policies. Now that you have an overview of the primary tabs in the Burp Suite interface, let's delve deeper into how to navigate and use these tools effectively. Within each tab, you'll find a set of sub-tabs, menus, and options tailored to the specific functionality of that tab. For example, in the Proxy tab, you can access sub-tabs for Intercept, HTTP history, and WebSockets, each providing different views and controls for managing web traffic. In the Spider tab, you can initiate and control the crawling process, view the site map, and configure various spidering options. Each sub-tab and feature within Burp Suite is designed to help you perform specific tasks related to web application security testing. To navigate between tabs and sub-tabs, you can simply click on the tab headers or use keyboard shortcuts, allowing for quick and efficient switching between different views and tools. As you work within the Burp Suite interface, you'll notice that it provides real-time feedback on the activity and results of your testing. For example, when intercepting requests in the Proxy tab, you'll see incoming traffic displayed in the Intercept sub-tab, where you can choose to forward, drop, or modify the requests before they reach the target web application. Similarly, when using the Scanner tab, you'll receive immediate feedback on vulnerabilities discovered during automated scans, helping you prioritize and address them promptly. The

Burp Suite interface also allows you to customize your workspace by arranging tabs, sub-tabs, and tool windows according to your preferences. You can drag and drop tabs to rearrange them, dock tool windows to different areas of the interface, and create custom layouts that suit your workflow. This flexibility ensures that you can optimize your workspace to focus on the tasks at hand. Furthermore, Burp Suite supports multi-tabbed browsing, allowing you to maintain separate sessions and configurations for different projects or assessments. You can easily switch between different contexts within the same instance of Burp Suite, streamlining your workflow and organization. As you navigate through the Burp Suite interface, you'll discover various features and functionalities that empower you to conduct thorough web application security assessments. From intercepting and modifying requests to scanning for vulnerabilities and automating attacks, Burp Suite offers a comprehensive toolkit for security professionals. It's important to explore each tab and tool, gaining familiarity with their capabilities and how they can be applied to different testing scenarios. Additionally, Burp Suite provides extensive documentation and resources to help you master its interface and make the most of its features. By becoming proficient in navigating and utilizing the Burp Suite interface, you'll be well-equipped to uncover vulnerabilities, assess web application security, and contribute to a more secure online environment. In the following chapters, we will dive deeper into each tab and tool within Burp Suite, providing practical insights *and examples to enhance your web application security testing skills. So, as you continue your journey with Burp Suite, embrace the power of its interface, and let it be your trusted companion in the quest for web application security.*

Chapter 3: Configuring Burp Suite for Your Needs

Proxy configuration and settings are fundamental aspects of using Burp Suite effectively for web application security testing. As mentioned earlier, Burp Suite acts as a proxy server, sitting between your browser and the target web application, capturing and analyzing all HTTP requests and responses. Configuring the proxy correctly is the first step in harnessing Burp Suite's power for intercepting and manipulating web traffic. To configure your browser to use Burp Suite as a proxy, you'll need to modify your browser's proxy settings. These settings typically include specifying the proxy host (which is often "localhost" or the IP address where Burp Suite is running) and the port number (by default, Burp Suite uses port 8080). Once your browser is configured to use Burp Suite as a proxy, all your web traffic will be routed through Burp Suite, allowing you to inspect, modify, and analyze the requests and responses in real-time. It's essential to ensure that your browser's proxy settings align with the configuration of Burp Suite to establish a seamless connection. In Burp Suite, you can configure the proxy settings by navigating to the "Proxy" tab, where you'll find a wide range of options to tailor the proxy behavior to your needs. One crucial setting is the "Bind to port" option, which specifies the port on which Burp Suite will listen for incoming traffic. The default port is 8080, but you can change it to any other available port if needed. Keep in mind that if you change the port in Burp Suite, you must also update your browser's proxy settings accordingly to match the new port. Additionally, Burp

Suite offers the option to use SOCKS proxy, which is particularly useful when testing applications that use non-HTTP protocols. You can configure SOCKS proxy settings in the "Options" menu under "Connections" in the "Upstream Proxy Servers" section. By default, Burp Suite listens on all network interfaces, but you can restrict it to listen only on specific interfaces if you have multiple network adapters. This can be useful in scenarios where you want to test applications on a specific network. The "Request handling" section in the proxy settings allows you to specify how Burp Suite should handle requests from your browser. You can choose to intercept requests, forward them directly, or drop them entirely. Interception is a valuable feature when you want to inspect and modify requests before they reach the target web application. It allows you to halt the request in the "Intercept" sub-tab of the "Proxy" tab, review its content, and decide whether to forward or drop it. This level of control is crucial for understanding how web applications work and for identifying vulnerabilities such as SQL injection and cross-site scripting (XSS). The "Intercept client requests" and "Intercept server responses" checkboxes let you specify which side of the communication you want to intercept. You can customize the interception rules by creating specific matching and scope rules, ensuring that only relevant traffic is intercepted. This is particularly useful when working with complex web applications or testing specific parts of a site. Burp Suite also allows you to configure proxy chaining, which means routing traffic through multiple proxy servers. This can be helpful when you need to access web applications that are behind additional layers of security or when you want to anonymize your testing traffic. In the

"Options" menu, under "Connections," you can configure proxy chaining settings by specifying the proxy host and port for each proxy server in the chain. By configuring proxy chaining, you can extend your testing capabilities and adapt to various testing scenarios. When using Burp Suite as a proxy, it's essential to consider the impact on your testing environment and ensure that you're not inadvertently sending testing traffic to production systems. To avoid this, you can configure scope settings in Burp Suite to define the domains and URLs that are within the scope of your testing. Scope settings are available in the "Target" tab, allowing you to specify the target's scope by adding domains, URL patterns, and exclusions. By defining a clear scope, you can ensure that only the intended web applications are tested, reducing the risk of unintended consequences. In addition to these proxy settings, Burp Suite offers advanced features such as upstream proxy support, proxy listeners for non-standard ports, and DNS resolution options. These advanced configurations can be valuable in complex testing environments or scenarios that require specific proxy behaviors. As you become more familiar with Burp Suite and its proxy capabilities, you'll gain a deeper understanding of how to configure and fine-tune the proxy settings to meet your specific testing needs. It's important to remember that while the default settings in Burp Suite are suitable for most testing scenarios, customizing the proxy configuration allows you to tailor Burp Suite to your unique requirements. Proxy configuration is a fundamental aspect of using Burp Suite effectively, enabling you to intercept and analyze web traffic between your browser and the target web application. By mastering the proxy settings and

understanding how to configure them to suit your testing goals, you'll unlock the full potential of Burp Suite as a web application security testing tool. In the next chapters, we will explore the practical use of the proxy in various testing scenarios and delve into advanced interception and manipulation techniques. As you continue your journey with Burp Suite, you'll discover how these proxy settings play a pivotal role in identifying vulnerabilities, assessing security, and contributing to a more secure web environment.

Target configuration and scope management are crucial aspects of web application security testing using Burp Suite. Next, we will delve into the significance of defining a clear target scope and how to configure it effectively. When conducting web application security assessments, it's essential to have a well-defined target scope to ensure that you focus your testing efforts on the right assets. Burp Suite provides tools and features to help you specify the domains, URLs, and applications that are within the scope of your testing. The "Target" tab is the central location for managing your target scope within Burp Suite. Here, you can add, modify, or remove target scope items to tailor your testing to the specific assets you want to assess. One of the first steps in configuring your target scope is to define the scope type: "Include in scope" or "Exclude from scope." "Include in scope" specifies that the defined items are the ones you want to test, while "Exclude from scope" implies that everything except the defined items is in scope. This flexibility allows you to choose the most suitable approach based on your testing objectives. Adding items to your target scope is straightforward. You can manually input domains or URL patterns, or you can use

Burp Suite's automated tools to discover and populate your scope. The "Add" button in the "Scope" tab lets you input domains and URL patterns manually. For example, if you want to test a specific web application hosted at "www.example.com," you can add "www.example.com" to your target scope. To test a broader range of URLs within that application, you can use wildcard patterns, such as "*.example.com," to include all subdomains. By defining your target scope with precision, you ensure that you are testing only what's necessary, reducing the risk of unintended consequences during your assessment. Burp Suite also provides features to automate the discovery and addition of items to your target scope. The "Spider" tool in Burp Suite is particularly useful for this purpose. It crawls the target web application, identifying and mapping its content and functionality, and automatically adds discovered items to your target scope. Using the "Spider" tool is beneficial when you have a large or complex web application to test. To configure and start the "Spider," navigate to the "Spider" tab in Burp Suite and specify the URL to begin crawling. You can customize spidering options to control the depth and breadth of the crawl, ensuring that it aligns with your testing objectives. As the "Spider" tool explores the web application, it adds discovered domains and URLs to your target scope, keeping it up to date with the application's structure. It's important to periodically review and refine your target scope as your assessment progresses. This ensures that you remain focused on testing the areas of the application that are relevant to your security goals. You can access the "Scope" tab in the "Target" tab to view and manage the items in your scope. Here, you can edit, remove, or clear *scope items as needed. You may encounter situations where you want to exclude specific items from your target scope. For example, if you discover a domain or URL that is not relevant to your assessment or leads to unintended consequences, you can exclude it. To exclude items, navigate to the "Scope" tab, select the item you want to exclude, and choose the "Exclude from scope" option. This ensures that the item is no longer considered within the scope of your testing. The "Scope" tab also provides the option to use regular expressions when specifying URL patterns, allowing for more advanced and flexible scope definitions. Regular expressions enable you to match URLs based on complex patterns, which can be particularly useful when testing applications with dynamic URLs. For example, you can use regular expressions to include all URLs that match a specific pattern, such as "/product/[0-9]+," where "[0-9]+" represents any sequence of digits. This approach allows you to encompass a broad range of URLs with a single pattern. Burp Suite also offers the ability to import and export your target scope configuration. This feature can be handy when collaborating with team members or when you want to reuse a specific scope configuration across multiple assessments. You can export your current scope configuration as a JSON or XML file and share it with others or save it for future use. To import a scope configuration, simply select the appropriate file in the "Scope" tab, and Burp Suite will apply the defined scope items. By effectively managing your target scope in Burp Suite, you ensure that your testing efforts are focused on the areas that matter most. This not only enhances the efficiency of your assessments but also reduces the risk of* unintentional impact on production systems. A well-defined target scope is a fundamental aspect of responsible and effective web application security testing. As you gain experience with Burp Suite and scope management, you'll develop a better understanding of how to align your testing with your specific objectives. Remember that maintaining an up-to-date and precise target scope is an ongoing process throughout your assessment. Periodic reviews and adjustments are essential to adapt to changing requirements and to ensure that your testing remains both effective and responsible. In the following chapters, we will explore how to use Burp Suite's various tools and features within your defined target scope to identify vulnerabilities, assess security, and contribute to a safer web environment. With a clear target scope in place, you are well-prepared to dive deeper into web application security testing using Burp Suite.

Chapter 4: Understanding HTTP and Web Protocols

Understanding the fundamentals of HTTP requests and responses is essential when conducting web application security testing with Burp Suite. HTTP, which stands for Hypertext Transfer Protocol, is the foundation of data communication on the World Wide Web. It is a protocol used by web browsers and servers to exchange information, making it possible for users to access and interact with web applications. HTTP operates on a client-server model, where the client, typically a web browser, sends requests to a server, and the server responds with the requested data. Each HTTP interaction consists of two primary components: the HTTP request and the HTTP response. The HTTP request is initiated by the client, and it specifies the action the client wants the server to perform. Common HTTP request methods include GET, POST, PUT, DELETE, and more, each serving a specific purpose. For example, the GET method is used to retrieve data from the server, while the POST method is used to send data to the server to be processed. HTTP requests also include a URL, which identifies the resource the client is requesting, such as a web page or an API endpoint. Additionally, requests may contain headers, which provide metadata about the request, and a message body, which carries data sent from the client to the server. Headers can convey information like the client's user agent, accepted content types, and cookies, among other details. The message body is optional and is often used with methods like POST to send data, such as form submissions or JSON payloads, to the server for processing. Once the server receives an HTTP request, it processes the request based on the provided method, URL, and any included data. The server then formulates an HTTP response to send back to the client. HTTP responses contain important information about the outcome of the request. One of the most critical components of an HTTP response is the status code, a three-digit numeric code that indicates whether the request was successful or encountered an error. Common status codes include 200 OK (indicating success), 404 Not Found (indicating the requested resource was not found), and 500 Internal Server Error (indicating a server-side error). HTTP responses also include headers, similar to requests, which provide metadata about the response. Headers can convey information such as the server's software version, content type, and caching instructions. The message body of an HTTP response typically contains the data that the client requested, such as an HTML page, an image, or JSON data. Understanding the structure and content of HTTP requests and responses is vital for web application security testing with Burp Suite. Burp Suite acts as a proxy between the client (usually your web browser) and the server, intercepting and analyzing these requests and responses. By examining and manipulating these interactions, you can identify security vulnerabilities and assess the overall security of web applications. When using Burp Suite's proxy functionality, you have the option to intercept and modify both HTTP requests and responses. Intercepting requests allows you to review and potentially modify the data being sent from your browser to the server. This capability is invaluable for understanding how web applications work and for identifying vulnerabilities such as SQL injection or cross-site scripting (XSS). For example, you can intercept a login request, review the parameters being sent (e.g., username and password), and manipulate them to test for authentication bypass vulnerabilities. Intercepting responses, on the other hand, allows you to inspect the data sent by the server back to your browser. This is crucial for identifying security misconfigurations or sensitive information disclosures in the server's responses. You can analyze the content of the responses, looking for clues about the web application's behavior or potential vulnerabilities. Burp Suite provides a user-friendly interface for intercepting and modifying HTTP requests and responses. In the "Proxy" tab, you can enable the interception feature, which allows you to pause requests and responses, review their content, and decide whether to forward or modify them before they reach the server or your browser. By actively engaging with these interactions, you gain insights into the application's security posture and can effectively test for vulnerabilities. Beyond interception, Burp Suite offers tools like the "Repeater" and the "Intruder" for further analysis and manipulation of HTTP requests. The "Repeater" tool allows you to send individual HTTP requests to the server multiple times, making it ideal for fine-tuning and verifying vulnerabilities. You can modify the request parameters and observe how the server responds to different inputs. The "Intruder" tool, on the other hand, is a powerful automation tool for conducting attacks on web applications. It enables you to define and execute customized attack scenarios by automatically varying input values and observing the application's responses. This is particularly useful for testing the security of input

fields and parameters. To summarize, understanding HTTP request and response basics is foundational to successful web application security testing with Burp Suite. HTTP interactions, composed of requests and responses, drive the communication between clients and servers on the web. Burp Suite's interception and manipulation capabilities allow you to examine and test these interactions effectively, helping you identify vulnerabilities and assess the security of web applications. In the upcoming chapters, we will delve deeper into using Burp Suite to intercept, modify, and analyze HTTP requests and responses in various security testing scenarios. By mastering these techniques, you'll be well-prepared to uncover vulnerabilities and contribute to enhancing web application security.
Web protocols are the building blocks of communication on the internet, serving as the foundation for data exchange between computers and devices. These protocols define the rules and conventions that enable the seamless transfer of information across the World Wide Web. Understanding web protocols and their significance is crucial for web developers, network administrators, and security professionals alike. One of the fundamental web protocols is HTTP, which stands for Hypertext Transfer Protocol. HTTP governs how data is requested and transmitted over the web, allowing users to access and interact with web pages and applications. HTTP is the protocol responsible for the familiar "http://" or "https://" prefix in website URLs. Another important web protocol is HTTPS, which stands for Hypertext Transfer Protocol Secure. HTTPS adds a layer of encryption to HTTP, ensuring that data transmitted between the client (usually a web

browser) and the server remains confidential and secure. This encryption is especially critical for safeguarding sensitive information like login credentials and personal data. HTTP and HTTPS are just two examples of the many web protocols that facilitate different aspects of web communication. SMTP, or Simple Mail Transfer Protocol, is used for sending and receiving email messages, while POP3 (Post Office Protocol 3) and IMAP (Internet Message Access Protocol) are used for retrieving emails from mail servers. FTP, or File Transfer Protocol, is employed for transferring files between computers on the internet. DNS, or Domain Name System, is responsible for translating human-readable domain names (like www.example.com) into IP addresses, allowing computers to locate each other on the network. These web protocols work together to enable various internet services, from web browsing and email to file sharing and online gaming. Their significance lies in their ability to standardize and streamline communication across the vast and interconnected web. When it comes to web application security, understanding web protocols is paramount. Security professionals use this knowledge to identify vulnerabilities and mitigate potential threats. One common security concern is the improper handling of input data in web applications. For instance, an application that fails to validate user input properly may be vulnerable to attacks like SQL injection or cross-site scripting (XSS). By understanding how web protocols work, security experts can manipulate input data to test whether an application is susceptible to these types of vulnerabilities. Web protocols also play a critical role in the secure transmission of data. HTTPS, in particular, is essential for protecting sensitive information. When you enter your credit card details on an e-commerce website, the data is encrypted using the HTTPS protocol, making it difficult for malicious actors to intercept and steal the information during transmission. Furthermore, web protocols are essential for authentication and authorization mechanisms in web applications. Protocols like OAuth and OpenID Connect enable users to log in securely to multiple websites using their existing credentials from trusted identity providers like Google or Facebook. This reduces the need for users to remember numerous usernames and passwords, enhancing both security and convenience. Web protocols are continually evolving to meet the growing demands of modern web applications and the need for increased security. For example, the development of HTTP/2 introduced performance improvements such as multiplexing and header compression, making web pages load faster. HTTP/3, built on the QUIC protocol, aims to further enhance web performance and security by reducing latency and improving encryption. As web protocols evolve, web developers and security professionals must stay up to date with the latest standards and best practices. By doing so, they can ensure that their applications are secure, efficient, and compliant with the latest industry standards. Moreover, web protocols also impact the user experience. For instance, the adoption of new protocols and technologies can enable the creation of interactive and dynamic web applications. WebSockets, a protocol that allows for real-time communication between a client and a server, has revolutionized web applications by enabling features like live chat, online gaming, and collaborative document editing. Web protocols are the *invisible infrastructure that underpins the modern internet, enabling everything from simple web browsing to complex cloud services. Their significance cannot be overstated, as they shape how information is transmitted, accessed, and secured on the web. Whether you're a developer building web applications, a network administrator managing web servers, or a security professional assessing the security of web assets, a solid understanding of web protocols is essential. It empowers you to harness the capabilities of these protocols effectively, ensuring the reliability, security, and performance of web services. In the ever-evolving landscape of web technologies, staying informed about the latest advancements and best practices in web protocols is a continuous journey. By embracing this journey, you can contribute to the growth and security of the internet while delivering better web experiences for users around the world.*

Chapter 5: Scanning and Crawling with Burp Suite

Web application crawling is a fundamental step in the process of web security testing, and understanding the techniques involved is crucial. Crawling is the process of systematically exploring a web application's structure, identifying its pages, and mapping its functionality. By doing so, security professionals can gain insights into the application's architecture and discover potential attack surfaces. One common crawling technique is the use of web spiders, which are automated tools that navigate through a web application by following links and recording the pages they encounter. These spiders start from a specific entry point, typically the application's homepage, and then traverse the links they find on each page. Web spiders, like Burp Suite's built-in spider, are efficient for discovering publicly accessible pages and resources within an application. They help create a site map that outlines the relationships between different pages and provides a clear overview of the application's structure. Web spiders can also follow forms on web pages and submit them, which allows them to discover hidden or dynamically generated content. However, web spiders have limitations. They may not be able to access pages that require user authentication, as they usually lack the necessary credentials. Additionally, they might not handle complex JavaScript-driven interactions that generate content dynamically. In such cases, manual intervention or other techniques may be necessary. Another technique for crawling web applications is manual exploration, where testers navigate through the application manually, following links and using the application's features as real users would. Manual exploration is valuable for testing user-specific functionality, such as authentication mechanisms and user profile pages. It also allows testers to interact with the application's user interface and analyze the behavior of JavaScript-driven features. By using manual exploration in combination with automated crawling, security professionals can ensure comprehensive coverage of the application. When dealing with complex single-page applications (SPAs) that rely heavily on JavaScript for content generation, traditional web spiders may fall short. In such cases, dynamic analysis tools, often referred to as headless browsers, can be employed. These tools, like Puppeteer and Playwright, provide a programmable interface to control web browsers programmatically. Security professionals can use headless browsers to interact with web applications, execute JavaScript, and retrieve dynamically generated content. By scripting interactions with the application, testers can explore SPAs thoroughly, just as a real user would. Additionally, some web security testing tools, including Burp Suite, offer features for crawling SPAs. Burp Suite, for example, includes a crawler with JavaScript rendering capabilities that can handle complex web applications. It uses headless browsers to execute JavaScript and explore SPAs effectively. An important consideration when crawling web applications is the scope. The scope defines which parts of the application are included in the crawl and which are excluded. Security testers need to define a clear scope to ensure that the crawl focuses on the areas of interest and avoids unintended consequences. For example, a tester may want to exclude third-party domains or administrative interfaces from the crawl to prevent accidental disruption or exposure of sensitive data. In Burp Suite, you can define the scope using the "Scope" tab in the "Target" tab. By specifying the target's scope, you can ensure that the crawler only explores the desired portions of the application. It's important to remember that web application crawling can be resource-intensive. Crawling large or complex applications can generate a significant amount of traffic and place a considerable load on the web server. To avoid causing disruption, testers should set appropriate crawl rate limits to control the speed at which the crawler operates. This ensures that the crawl remains respectful of the application's resources and doesn't impact its availability to users. The results of web application crawling provide a foundation for subsequent phases of security testing. Once the crawl is complete, testers can use the discovered URLs and pages as targets for further analysis. For example, they can use the "Scanner" tool in Burp Suite to automatically test the identified pages for common vulnerabilities such as SQL injection and cross-site scripting (XSS). The site map generated during crawling serves as a roadmap for testing and helps testers keep track of their progress. In summary, web application crawling is a critical step in the process of web security testing. By systematically exploring an application's structure and mapping its functionality, testers can identify potential attack surfaces and assess its security posture. Automated web spiders, manual exploration, and headless browsers are among the techniques used for crawling web applications. Defining a clear scope and setting appropriate crawl rate limits are essential considerations to ensure effective and responsible crawling. The results of the crawl serve as the basis for further vulnerability assessment and testing phases. Overall, web application crawling is a key component of the comprehensive security testing process, helping to uncover vulnerabilities and contribute to a more secure online environment.

Automated vulnerability scanning is a vital component of modern web application security testing, providing a systematic and efficient way to identify potential security issues. In today's digital landscape, where new web applications are constantly being developed and deployed, security professionals face the daunting task of continuously assessing and securing these applications. Manual testing alone is often impractical due to time constraints and the sheer volume of web assets. This is where automated vulnerability scanning tools come into play, offering a scalable solution to help security teams keep pace with the evolving threat landscape. Automated vulnerability scanners are designed to mimic the actions of a human tester, systematically crawling a web application, sending requests, and analyzing responses to identify security *vulnerabilities. These tools can save time and effort by automatically detecting common security issues, such as SQL injection, cross-site scripting (XSS), and insecure authentication mechanisms. One of the key advantages of automated vulnerability scanning is its speed and efficiency. Scanning tools can assess a large number of web pages and assets in a relatively short amount of time, making them indispensable for organizations with extensive web application portfolios. Automated scans can be scheduled to run regularly, ensuring continuous monitoring of web applications for new vulnerabilities or changes that may introduce security risks. Furthermore, automated scanning tools can operate 24/7, providing round-the-clock security coverage. This continuous scanning approach helps organizations detect and remediate vulnerabilities promptly, reducing the window of opportunity for potential attackers. Another benefit of automated scanning is its ability to provide consistent and repeatable results. Unlike manual testing, which can be influenced by human error or bias, automated scans follow predefined rules and testing methodologies consistently. This consistency is essential for accurately identifying vulnerabilities and assessing their severity. Automated vulnerability scanners often produce detailed reports that include information about each discovered issue, including its location, severity, and a description of the vulnerability. These reports are valuable for security teams and developers, as they provide actionable insights into the security posture of the web application. Security professionals can prioritize remediation efforts based on the severity of the* vulnerabilities identified in the scan reports. Additionally, automated scanning tools often provide proof of concept (PoC) payloads that demonstrate the exploitability of vulnerabilities, helping security teams understand the potential impact. While automated vulnerability scanning offers numerous advantages, it's essential to recognize its limitations. No scanning tool can replace the expertise of a skilled human tester who can perform in-depth assessments and identify complex vulnerabilities that automated tools may miss. Automated scanners may produce false positives, indicating vulnerabilities that do not actually exist, or false negatives, failing to detect certain vulnerabilities. To address these limitations, security teams should combine automated scanning with manual testing to achieve comprehensive coverage. Automated scanning tools are most effective when they are integrated into the development and deployment pipeline. This approach, often referred to as "shift-left" security, involves conducting automated scans at various stages of the software development lifecycle. By identifying and addressing vulnerabilities early in the development process, organizations can reduce the cost and effort required to remediate issues. Automated scanning can be integrated into continuous integration/continuous deployment (CI/CD) pipelines, allowing security checks to occur automatically as new code is built and deployed. This ensures that vulnerabilities are caught early, minimizing the risk of deploying insecure code into production. Additionally, automated scans can be used to validate the security of third-party components and *libraries used in web applications. These components can introduce vulnerabilities, so it's crucial to regularly scan them for potential issues. Automated scanning can also assist organizations in complying with regulatory requirements and industry standards. Many compliance frameworks require regular security assessments, and automated scans can help organizations demonstrate their commitment to security. When selecting an automated vulnerability scanning tool, organizations should consider factors such as the tool's coverage of different vulnerability types, its accuracy, and its ability to integrate with existing security and development workflows. Some popular automated scanning tools include Burp Suite's scanner module, OWASP ZAP, Nessus, Qualys, and Acunetix, among others. Ultimately, the choice of tool should align with the organization's specific needs and goals. In summary, automated vulnerability scanning is a critical component of modern web application security testing. These tools offer speed, efficiency, and consistency in identifying security vulnerabilities, enabling organizations to assess and mitigate risks effectively. However, it's essential to recognize the limitations of automated scanning and complement it with manual testing for comprehensive coverage. Integrating automated scanning into the development and deployment pipeline can help catch vulnerabilities early, reducing security risks and ensuring the delivery of secure web applications. By leveraging automated scanning as part of a holistic security strategy, organizations can enhance their ability to protect against evolving threats in the digital landscape.*

Chapter 6: Intercepting and Modifying Web Requests

Intercepting and analyzing requests is a fundamental technique in web application security testing, allowing security professionals to gain insight into the data sent between a client and a server. By intercepting requests, testers can examine the information exchanged, including parameters, headers, and cookies, to identify security vulnerabilities and assess the overall security of web applications. Interception is a core feature of security testing tools like Burp Suite, providing testers with the ability to inspect, manipulate, and analyze HTTP requests before they reach the server. To get started with intercepting requests in Burp Suite, you first configure your browser to use Burp Suite as a proxy. This means that all HTTP traffic generated by your browser will be routed through Burp Suite, giving you full control over the requests and responses. Once your browser is configured to use Burp Suite as a proxy, you can begin intercepting requests by enabling the interception feature in Burp Suite's proxy tab. When interception is enabled, Burp Suite will pause incoming requests from your browser, allowing you to review and decide whether to forward or modify them before they are sent to the web server. This interception capability is invaluable for understanding how web applications work and for identifying vulnerabilities such as SQL injection or cross-site scripting (XSS). For example, you can intercept a login request, review the parameters being sent (e.g., username and password), and manipulate them to test for authentication bypass vulnerabilities. You can also analyze request headers to identify potential security misconfigurations, such as the absence of security headers like Content Security Policy (CSP) or Cross-Origin Resource Sharing (CORS) headers. Intercepting and analyzing requests goes beyond the surface of web applications, allowing testers to dig deep into their functionality. One common scenario is testing for insecure direct object references (IDOR), where an attacker can manipulate parameters to access unauthorized data. By intercepting requests, testers can modify parameters to access resources they shouldn't have access to and verify whether the application enforces proper authorization. Intercepting requests also provides a way to test input validation and sanitization. Testers can submit payloads designed to trigger common security vulnerabilities, such as SQL injection or XSS, and observe how the application processes the data. Analyzing the responses received after submitting such payloads can reveal whether the application is vulnerable to these attacks. Beyond security testing, intercepting and analyzing requests can be a valuable tool for understanding web application behavior. By examining the requests generated during various interactions with the application, testers can gain insights into its internal logic, the flow of data, and potential areas where security vulnerabilities may exist. Interception allows for real-time feedback, enabling testers to iteratively test and refine their understanding of the application. In addition to intercepting and reviewing individual requests, Burp Suite provides tools for saving and organizing intercepted requests and responses. Testers can use the "Proxy History" tab to view a chronological list of intercepted requests and responses, making it easy to navigate and reference past interactions. The "Proxy History" tab also includes options to search, filter, and tag intercepted items, aiding in the organization of test data. One of the advanced features of Burp Suite is its ability to automate the analysis of intercepted requests. Testers can create custom rules to automatically flag or manipulate requests based on specific criteria. For example, you can create a rule to highlight requests with certain keywords in their parameters or responses, making it easier to spot potential security issues. Burp Suite also offers extensibility through its support for user-written macros and extensions. Testers can develop custom macros and extensions to automate complex testing scenarios, enabling the tool to handle repetitive tasks and accelerate the testing process. For example, you can create a custom extension to automatically generate and inject payloads for common security vulnerabilities like SQL injection or XSS. Intercepting and analyzing requests is a foundational skill for web application security testing, and Burp Suite provides a user-friendly interface to support this essential task. By mastering the interception capabilities of Burp Suite and combining them with other testing techniques, security professionals can effectively assess the security of web applications and contribute to making them more resilient to attacks. In the following chapters, we will explore additional techniques and tools that can be used in conjunction with request interception to uncover vulnerabilities and enhance web application security.

Modifying requests for testing is a crucial aspect of web application security testing, allowing security professionals to assess how an application responds to different inputs and scenarios. By altering the data sent in HTTP requests, testers can identify vulnerabilities, uncover security misconfigurations, and evaluate the robustness of the application's defenses. Modifying requests gives testers the ability to simulate various attack vectors and explore the application's behavior under different conditions. In security testing, the objective is to identify and remediate vulnerabilities before malicious actors can exploit them. One common way to modify requests is to manipulate input parameters. Input parameters are values provided to the application through forms, URLs, or headers. Testers can modify these parameters to test for security vulnerabilities such as SQL injection, cross-site scripting (XSS), and insecure direct object references (IDOR). For example, in a login form, testers can modify the username and password fields to check if the application is vulnerable to authentication bypass. By injecting special characters or payloads designed to trigger vulnerabilities, testers can observe how the application processes the modified data. Another approach to modifying requests is to manipulate headers, which provide additional information about the request or the client. Headers like "User-Agent" and "Referer" can be changed to simulate requests from different browsers or sources. This can help testers assess whether the application handles different user agents or referral sources correctly and doesn't rely on them for security. In some cases, modifying headers can be used to bypass security controls that are based on specific headers. For example, security headers like Content Security Policy (CSP) or Cross-Origin Resource Sharing (CORS) can be tested by modifying the "Origin" header to see how the application responds. Cookies, *another type of data sent in headers, can also be manipulated to test for vulnerabilities. Testers can modify cookies to check if the application properly enforces session management and access control. By tampering with session cookies, testers can attempt to hijack sessions or impersonate other users. Furthermore, requests can be modified to test for issues related to file uploads. By altering the content of file uploads or changing file extensions, testers can assess whether the application adequately validates and processes uploaded files. This can help uncover vulnerabilities such as file inclusion or file execution vulnerabilities. While modifying requests for testing is a powerful technique, it's essential to conduct such testing responsibly and ethically. Testing should only be performed on applications or systems for which you have explicit authorization. Unauthorized testing can lead to legal consequences and may disrupt the availability and integrity of the application. Additionally, testers should be cautious when using automated tools to modify requests, as these tools can generate a high volume of traffic and potentially overload the application. Properly configuring the tools and setting rate limits can help prevent unintended disruption. In the context of web application security testing, tools like Burp Suite offer powerful capabilities for request modification. Burp Suite provides an intercepting proxy that allows testers to pause and modify requests in real-time before they are sent to the server. The "Repeater" tool in Burp Suite enables testers to send individual requests to the server multiple times, making it ideal for fine-tuning and verifying vulnerabilities. Testers can modify request parameters, headers, and cookies in the Repeater tool and observe how the server responds. Additionally, Burp Suite's extensibility allows testers to create custom macros and extensions to automate the modification of requests based on specific criteria or testing scenarios. When modifying requests, it's essential to document the changes made and the observed behavior of the application. Detailed notes and logs help testers keep track of their testing progress and share findings with development and security teams for remediation. Testers should also maintain a clear understanding of the application's functionality and logic to make informed modifications. In summary, modifying requests for testing is a vital technique in web application security testing. It empowers testers to simulate various attack scenarios, evaluate security controls, and identify vulnerabilities. However, testing should always be conducted responsibly and ethically, with proper authorization and safeguards to prevent unintended disruption. Tools like Burp Suite provide robust features for request modification and are essential for effective security testing. By mastering the art of request modification, security professionals can contribute to improving the security of web applications and protecting them from potential threats.*

Chapter 7: Analyzing and Exploiting Vulnerabilities

Identifying and assessing vulnerabilities is a critical step in web application security testing, helping security professionals understand and mitigate potential risks. Vulnerabilities are weaknesses or flaws in an application's design, code, or configuration that can be exploited by attackers to compromise its security. The goal of vulnerability assessment is to uncover these weaknesses, evaluate their impact, and prioritize them for remediation. One common approach to identifying vulnerabilities is to use automated scanning tools like Burp Suite's scanner module, which can scan web applications for known vulnerabilities. Automated scanners simulate attack scenarios and test web applications for issues such as SQL injection, cross-site scripting (XSS), and security misconfigurations. These tools can scan a large number of pages and assets quickly, making them valuable for initial assessments and continuous monitoring. However, automated scanners have limitations and may produce false positives or false negatives. A false positive occurs when a scanner incorrectly identifies a vulnerability that does not exist, while a false negative occurs when it fails to detect a real vulnerability. To address these limitations, manual testing is essential. Manual testing involves security professionals actively examining the application, its source code, and its behavior to identify vulnerabilities that automated tools may miss. During manual testing, testers explore the application as an attacker would, trying to exploit weaknesses in its defenses. For example, testers may attempt to manipulate input parameters, bypass authentication mechanisms, or access unauthorized data. Manual testing can uncover complex vulnerabilities and security issues that require human expertise to identify. Security professionals often follow testing methodologies such as the OWASP Testing Guide to ensure comprehensive coverage. Vulnerability assessment also involves reviewing and analyzing the results of automated scans. Testers need to validate the findings of automated scanners, prioritize vulnerabilities based on their severity, and provide actionable information to developers and system administrators for remediation. The Common Vulnerability Scoring System (CVSS) is a widely used framework for assessing the severity of vulnerabilities. CVSS assigns a numerical score to each vulnerability, considering factors like exploitability, impact, and the level of access required. This score helps prioritize vulnerabilities and allocate resources for remediation effectively. Another aspect of vulnerability assessment is identifying and assessing the potential impact of vulnerabilities. Testers need to determine what an attacker could achieve if they successfully exploit a vulnerability. For example, an SQL injection vulnerability could allow an attacker to access, modify, or delete sensitive data in a database. Understanding the potential impact helps organizations assess the risk associated with each vulnerability and make informed decisions about mitigation. Additionally, vulnerability assessment includes considering the context in which an application operates. External factors such as the application's environment, the type of data it handles, and its user base can influence the severity of vulnerabilities. For instance, a vulnerability in a public-facing e-commerce website that handles financial transactions is typically more critical than a vulnerability in an internal employee portal. Testers need to take these contextual factors into account when assessing vulnerabilities and prioritizing remediation efforts. In addition to traditional web applications, vulnerability assessment should also include APIs (Application Programming Interfaces) and web services. APIs are a common target for attackers, as they often expose critical functionality and sensitive data. Security professionals need to assess the security of APIs, checking for vulnerabilities like insecure authentication, broken access control, and data exposure. Furthermore, vulnerability assessment is an ongoing process that should be integrated into the software development lifecycle. Security should be a consideration at every stage of application development, from design and coding to testing and deployment. By incorporating security practices early in the development process, organizations can reduce the likelihood of vulnerabilities being introduced and minimize the cost of remediation. Security testing should be conducted regularly, especially when changes are made to the application or its environment. Regular testing helps organizations identify and address new vulnerabilities that may emerge over time. Penetration testing, which involves simulating real-world attacks on the application, is another valuable technique for identifying vulnerabilities. Penetration testers use their skills and knowledge to exploit weaknesses in the application's defenses and provide detailed reports of their findings. These reports guide organizations in prioritizing and remediating vulnerabilities effectively. Vulnerability assessment tools and platforms, like Burp Suite, offer features for managing and tracking vulnerabilities throughout the remediation process. They enable testers to collaborate with developers and track the status of vulnerabilities from discovery to resolution. In summary, identifying and assessing vulnerabilities is a critical aspect of web application security testing. It involves a combination of automated scanning, manual testing, and analysis to uncover weaknesses in an application's security. Vulnerability assessment considers factors such as severity, impact, and context to prioritize vulnerabilities for remediation. It should be an integral part of the software development lifecycle, conducted regularly to ensure the ongoing security of web applications and APIs. By proactively identifying and addressing vulnerabilities, organizations can enhance the security of their web applications and protect them from potential threats. Exploiting vulnerabilities for testing purposes is a critical component of web application security testing, enabling security professionals to verify the presence of vulnerabilities and assess their real-world impact. This testing approach involves deliberately exploiting known or suspected vulnerabilities within an application to understand how they can be leveraged by attackers. The goal is to simulate the actions of a malicious actor and evaluate the extent of potential harm that can result from these vulnerabilities. Exploiting vulnerabilities allows testers to validate the accuracy of their findings, verify the severity of the vulnerabilities, and provide concrete evidence to developers and stakeholders for remediation. It is important to note that this activity should only be conducted with explicit authorization and in a controlled, isolated testing environment to prevent harm to production systems and data. One common type of vulnerability that testers often exploit is SQL injection. SQL injection occurs when an attacker can manipulate input data to inject malicious SQL queries into an application's database. To test for SQL injection vulnerabilities, testers deliberately craft input values that include SQL statements, such as UNION-based or blind SQL injection payloads, and submit them to the application. By observing the application's response, testers can determine whether the vulnerability exists and assess its impact. For example, exploiting an SQL injection vulnerability may allow testers to retrieve sensitive data from the database or even modify its contents. Another frequently tested vulnerability is cross-site scripting (XSS), which involves injecting malicious scripts into web pages viewed by other users. To exploit XSS vulnerabilities, testers insert scripts, typically JavaScript, into input fields or parameters that are reflected in web page content. Upon successful exploitation, the malicious script executes within the context of other users' browsers, potentially stealing their session cookies or performing other malicious actions. Exploiting XSS vulnerabilities demonstrates the risk of client-side attacks and helps prioritize remediation efforts. Insecure authentication mechanisms are also targets for exploitation. Testers may attempt to bypass authentication by using known usernames and passwords or by exploiting flaws in the login process, such as weak session management. By successfully bypassing authentication, testers can access unauthorized areas of the application or impersonate other users. Testing for broken access control involves exploiting vulnerabilities related to access permissions. Testers may manipulate URLs, parameters, or cookies to access restricted functionality or data. Exploiting such vulnerabilities helps identify areas where proper access controls are not enforced and can lead to unauthorized data exposure or modification. Web application firewalls (WAFs) are security controls designed to protect web applications from various attacks. To assess the effectiveness of a WAF, testers may attempt to evade or bypass it by using techniques like encoding, obfuscation, or evasion. By successfully bypassing the WAF, testers can highlight areas where additional security controls or rule tuning may be necessary. Exploiting vulnerabilities related to session management is also a critical aspect of testing. Testers may attempt to hijack sessions by stealing session cookies or tokens. Session hijacking can lead to unauthorized access to user accounts and sensitive data. Understanding the impact of session-related vulnerabilities is essential for ensuring user data and privacy. In addition to exploiting known vulnerabilities, testers often explore the *application for zero-day vulnerabilities—previously unknown issues that may exist. This involves a deeper level of analysis and creativity, as testers need to identify and exploit vulnerabilities that have not been documented or patched. Discovering zero-day vulnerabilities requires a high level of expertise and is typically performed by skilled security researchers and penetration testers. In the context of web application security testing, it is crucial to document and report the steps taken to exploit vulnerabilities, along with the observed outcomes. Comprehensive and detailed reports provide developers and stakeholders with a clear understanding of the risks and potential impact associated with each vulnerability. These reports serve as actionable insights for prioritizing and remediating vulnerabilities effectively. Exploiting vulnerabilities should always be conducted responsibly and ethically. Testers should obtain explicit authorization to perform testing activities, and all testing should be limited to a controlled testing environment to avoid harm to production systems and data. Additionally, testers should exercise caution when conducting tests that involve potentially destructive actions or sensitive data access. It is essential to minimize the impact of testing on the application and its users. In summary, exploiting vulnerabilities for testing purposes is a crucial step in web application security testing. It allows security professionals to validate the presence and impact of vulnerabilities, providing concrete evidence for remediation efforts. Testers target various vulnerabilities, including SQL injection, cross-site scripting (XSS), insecure authentication, broken access control, web application firewall (WAF) evasion, and session-related flaws. Ethical and responsible testing practices, including obtaining authorization and conducting tests in controlled environments, are essential to ensure the security of web applications and protect against potential threats.*

Chapter 8: Web Application Authentication Testing

Authentication mechanisms are a fundamental aspect of web application security, serving as the first line of defense against unauthorized access. These mechanisms verify the identity of users and determine whether they have the appropriate permissions to access the application's resources. Authentication plays a crucial role in safeguarding sensitive data and functionality. Effective testing of authentication mechanisms is essential to identify vulnerabilities and ensure that user access is properly controlled. Testing approaches for authentication mechanisms encompass a range of techniques designed to assess the security and reliability of the authentication process. One common area of focus in authentication testing is password security. Password-based authentication is widely used, and its security relies on factors such as password complexity, encryption, and storage. Testers may assess the strength of password policies by attempting to create accounts with weak or easily guessable passwords. Additionally, they may test for vulnerabilities like password brute-forcing, where an attacker tries various password combinations to gain unauthorized access. Another aspect of authentication testing is the assessment of multi-factor authentication (MFA) implementations. MFA enhances security by requiring users to provide two or more factors to authenticate, typically something they know (e.g., a password) and something they have (e.g., a mobile device). Testers can *assess whether MFA is properly implemented and whether it effectively mitigates the risk of unauthorized access. Authentication tokens, which are used to prove a user's identity, are subject to testing as well. Testers may attempt to forge or intercept authentication tokens to gain unauthorized access to user accounts. Testing for session management vulnerabilities is another critical aspect of authentication assessment. Session management ensures that users remain authenticated during their interactions with the application. Testers may explore session fixation, session hijacking, and session timeout vulnerabilities to assess the effectiveness of session management mechanisms. Authentication bypass testing involves attempting to access restricted areas or functionality without providing valid authentication credentials. Testers may manipulate input parameters, cookies, or URLs to bypass authentication and assess whether the application enforces access controls correctly. Biometric authentication methods, such as fingerprint or facial recognition, are becoming increasingly popular. Testing these methods involves evaluating their accuracy and resistance to spoofing or impersonation. Biometric authentication testing requires specialized tools and expertise. API security is a critical consideration in authentication testing, as APIs often expose authentication-related functionality. Testers assess API endpoints responsible for authentication, including their access controls, rate limiting, and error handling. Understanding the various testing approaches for authentication mechanisms is crucial for comprehensive* security testing. The OWASP Testing Guide and the OWASP Application Security Verification Standard provide valuable resources and methodologies for authentication testing. It's important to note that the goal of authentication testing is not only to identify vulnerabilities but also to provide recommendations for remediation. Effective testing should result in actionable insights that guide developers and administrators in improving authentication security. Furthermore, authentication testing should be performed throughout the software development lifecycle, from design and development to deployment and maintenance. This ensures that security considerations are integrated into the development process and that vulnerabilities are addressed promptly. In addition to manual testing, automated testing tools can assist in the assessment of authentication mechanisms. These tools can simulate various authentication scenarios, including password attacks, token manipulation, and session management tests. Burp Suite, for example, offers features for automated testing of authentication mechanisms, allowing testers to identify vulnerabilities more efficiently. As organizations transition to modern authentication methods, such as single sign-on (SSO) and identity federation, testing these mechanisms becomes crucial. SSO allows users to access multiple applications with a single set of credentials, streamlining the authentication process. Testing SSO implementations involves evaluating the trust relationships between identity providers and service providers, as well as assessing the security of SSO tokens and assertions. Identity federation extends SSO across multiple organizations, enabling users to access resources from different domains seamlessly. Testing identity federation involves verifying the security and trustworthiness of the federation protocols and configurations. Moreover, testing the integration of external identity providers and their impact on authentication is essential. In summary, authentication mechanisms are a fundamental element of web application security. Testing these mechanisms is crucial to ensure that user access is properly controlled, and vulnerabilities are identified and remediated. Authentication testing encompasses various techniques, including password security assessment, multi-factor authentication testing, session management testing, and API security assessment. Testing should be conducted throughout the development lifecycle, and automated testing tools can aid in the process. As authentication methods evolve, staying current with testing approaches is essential to maintain the security of web applications and protect against unauthorized access. Session management is a critical component of web application security, ensuring that users remain authenticated and authorized during their interactions with the application. However, session management vulnerabilities can lead to unauthorized access, allowing attackers to impersonate legitimate users. Understanding session management and how authentication bypass techniques can exploit weaknesses in the process is essential for effective security testing. Session management involves the creation, maintenance, and termination of user sessions, which are temporary states that allow users to access an application while logged in. Sessions are established after successful authentication, and they typically include a session identifier, which is a unique token used to associate requests with a specific session. One common session management vulnerability is session fixation, which occurs when an attacker sets the session identifier for a victim user before they log in. This means that the attacker can control the user's session, potentially gaining unauthorized access to their account. Session fixation attacks often involve tricking users into clicking on a malicious link that sets the session identifier. Authentication bypass, on the other hand, refers to techniques that allow attackers to access restricted areas or functionality without providing valid authentication credentials. Session management and authentication bypass techniques often go hand in hand, as session management vulnerabilities can be exploited to bypass authentication controls. One approach to testing session management is to assess how sessions are created and managed. Testers may attempt to manipulate session identifiers to gain unauthorized access to other users' sessions or escalate their privileges. For example, they can try to guess or brute-force session identifiers, hoping to land on a valid one. Additionally, testers may explore session fixation vulnerabilities by setting a victim user's session identifier and then attempting to use that session. Another aspect of session management testing involves assessing session timeouts. Session timeout vulnerabilities can lead to sessions remaining active even after a user logs out or leaves the application. Testers may investigate whether sessions persist longer than intended, potentially allowing attackers to hijack abandoned sessions. Effective session management also includes protecting sessions from being hijacked or stolen. Testing the resistance of session identifiers to theft or prediction is crucial. Testers may attempt to capture or predict session identifiers to gain unauthorized access. This can involve techniques like session hijacking, where attackers steal a user's session identifier, or session fixation, where attackers set a user's session identifier to a known value. Session management testing should also cover scenarios where users log out. Logging out should invalidate the user's session and prevent any further access to restricted areas or functionality. Testers may assess whether the logout process effectively terminates sessions and ensures that users cannot return to previously authenticated states. Furthermore, session management testing should consider the impact of network-level attacks, such as man-in-the-middle (MitM) attacks. MitM attackers may intercept and manipulate session identifiers during transmission, potentially compromising the security of sessions. Testers can simulate MitM attacks to assess how session identifiers are protected during transit and whether encryption and secure communication protocols are properly implemented. In addition to manual testing, automated tools can assist in identifying session management vulnerabilities. Tools like Burp Suite can automatically scan for session fixation, session timeout, *and session identifier issues. Testers can use these tools to accelerate the discovery of potential vulnerabilities. However, it's important to note that automated tools may not catch all session management issues, and manual testing remains essential for comprehensive assessment. In summary, session management is a critical component of web application security, ensuring that users remain authenticated and authorized. Session management vulnerabilities, such as session fixation and timeout issues, can lead to unauthorized access. Testing session management involves assessing how sessions are created, managed, and terminated, as well as evaluating their resistance to theft or prediction. Authentication bypass techniques often exploit session management weaknesses to gain unauthorized access. Effective security testing should encompass both session management and authentication bypass assessments to identify and remediate vulnerabilities. By understanding the intricacies of session management and authentication bypass, security professionals can contribute to the overall security of web applications and protect against unauthorized access.*

Chapter 9: Advanced Burp Suite Techniques

Customizing Burp Suite with extensions is a powerful way to enhance the functionality and capabilities of this versatile web application security testing tool. Extensions, also known as plugins or add-ons, allow security professionals to tailor Burp Suite to their specific needs and requirements. By customizing Burp Suite with extensions, testers can automate tasks, integrate with other tools, and uncover unique security vulnerabilities. Extensions in Burp Suite can be written in various programming languages, including Java, Python, and Ruby. This flexibility enables developers to create extensions that match their preferred language or skill set. The Burp Extender tool, provided by Burp Suite, serves as the platform for managing and running extensions. It offers an integrated development environment (IDE) for writing, testing, and deploying extensions. To get started with customizing Burp Suite using extensions, testers need to understand the different types of extensions available. There are three main categories of Burp Suite extensions: Burp Extender, Burp Scanner, and Burp Collaborator. Burp Extender extensions are the most versatile and allow testers to customize nearly every aspect of Burp Suite. They can add new functionality, modify existing features, and integrate with external tools and services. For example, an extension could be created to automate repetitive tasks, such as sending payloads to the Intruder tool for *fuzz testing. Burp Scanner extensions focus on enhancing the capabilities of the scanner module in Burp Suite. They can add custom checks for specific vulnerabilities, such as testing for authentication bypass or business logic flaws. These extensions expand the range of security assessments that Burp Suite can perform. Burp Collaborator extensions interact with the Burp Collaborator service, which is used to detect out-of-band vulnerabilities. These extensions can be used to automate interactions with the Burp Collaborator service, helping testers identify blind vulnerabilities, such as blind SQL injection or server-side request forgery (SSRF). Developing a Burp Suite extension begins with setting up a development environment. Testers need to install the Burp Extender tool and the necessary software development kits (SDKs) for their chosen programming language. Once the environment is set up, developers can start writing and testing their extension code. The Burp Extender IDE provides a rich set of tools for developing extensions, including syntax highlighting, code completion, and debugging capabilities. Extensions can be written from scratch or based on existing templates provided by PortSwigger, the company behind Burp Suite. These templates help developers get started quickly and provide a solid foundation for building extensions. When developing an extension, testers should have a clear understanding of their goals and objectives. They should define the specific functionality they want to add or modify in Burp Suite and plan the extension accordingly. Developers can leverage the Burp Extender API, which offers a wide range of functions and* interfaces for interacting with Burp Suite's core features. The API allows developers to access and manipulate HTTP requests and responses, modify scanner checks, and customize user interface components. Additionally, Burp Suite provides detailed documentation and examples to help developers understand and use the API effectively. Testing and debugging an extension is a crucial part of the development process. Developers should thoroughly test their extension to ensure it behaves as expected and does not introduce any errors or vulnerabilities into Burp Suite. The Burp Extender IDE includes tools for testing extensions within the Burp Suite environment. Developers can use the built-in debugger to step through their code, inspect variables, and identify issues. Once the extension is tested and verified, it can be packaged and deployed in Burp Suite. Extensions are distributed as JAR (Java Archive) files, which can be loaded into Burp Suite via the Extender tool. Developers can specify when and how the extension should be invoked, such as during a specific phase of the testing process or for particular URLs. Burp Suite provides flexibility in configuring the extension's behavior to meet the testing requirements. After deploying an extension, testers can use it to automate tasks, extend Burp Suite's functionality, and customize their security testing workflows. Extensions can save time and effort by automating repetitive tasks, allowing testers to focus on identifying and addressing security vulnerabilities. For example, an extension could be created to automatically detect and report potential security misconfigurations in web applications. Another common use case for extensions is integrating Burp Suite with other security tools and services. Testers can develop extensions that communicate with external scanners, vulnerability management systems, or threat intelligence feeds, streamlining the security assessment process. Additionally, extensions can enhance the reporting capabilities of Burp Suite by generating customized reports that highlight specific findings and remediation recommendations. It's important to note that while extensions can provide valuable enhancements to Burp Suite, they should be used responsibly and ethically. Testers should obtain proper authorization before using extensions in a testing environment and ensure that they comply with relevant laws and regulations. In summary, customizing Burp Suite with extensions is a powerful way to tailor the tool to specific testing needs and objectives. Extensions can be developed to automate tasks, add new functionality, and integrate with external tools and services. The Burp Extender tool and its integrated development environment make it accessible for security professionals to create and deploy extensions effectively. By harnessing the flexibility and extensibility of Burp Suite, testers can enhance their security testing capabilities and contribute to the overall security of web applications. Automation and scripting with Burp Suite are essential skills for security professionals, enabling them to streamline and optimize their web application security testing processes. While manual testing is valuable, automation can significantly increase efficiency, allowing testers to cover more ground and *identify vulnerabilities more quickly. Burp Suite provides a range of automation and scripting capabilities through its extensibility and powerful API. Automation in Burp Suite involves creating scripts or extensions that perform specific tasks or sequences of actions automatically. These scripts can interact with Burp Suite's core features, manipulate HTTP requests and responses, and even integrate with external tools and services. Automation can be applied to various aspects of web application security testing, from scanning and crawling to data analysis and reporting. One of the primary use cases for automation in Burp Suite is vulnerability scanning. Automated scanners can crawl web applications, identify vulnerabilities, and generate detailed reports. By automating the scanning process, testers can quickly identify common vulnerabilities such as SQL injection, cross-site scripting (XSS), and security misconfigurations. Burp Suite's built-in scanner module provides an easy way to automate vulnerability scanning, and testers can further customize and extend its capabilities using scripting. Scripting allows testers to fine-tune scanning parameters, create custom scan checks, and adapt the scanner to the specific needs of the application under test. For example, testers can script custom checks to detect business logic flaws or authentication bypass vulnerabilities that automated scanners may miss. In addition to vulnerability scanning, automation can be applied to other tasks, such as data extraction and manipulation. Testers can use scripts to extract data from HTTP responses, perform data-driven testing, and simulate different user roles or personas.*

For example, a script can automate the submission of various payloads to test input validation or perform brute-force attacks to identify weak credentials. Burp Suite's extensibility allows testers to create custom tools and utilities tailored to their testing requirements. Another critical aspect of automation is workflow automation. Testers can automate repetitive testing workflows, such as the process of identifying, verifying, and documenting vulnerabilities. By scripting these workflows, testers can ensure consistency in their testing practices and reduce the likelihood of overlooking critical issues. Automation scripts can also assist in post-exploitation activities, such as privilege escalation and data exfiltration. For example, a script can automate the process of escalating privileges in a web application by exploiting identified vulnerabilities. This can include automating actions such as gaining administrative access, extracting sensitive data, or executing commands on the underlying system. Burp Suite's scripting capabilities are not limited to vulnerability scanning and exploitation. Testers can automate the generation of reports and documentation, making it easier to communicate findings to stakeholders and developers. Custom report generation scripts can format scan results, highlight critical vulnerabilities, and include recommended remediation steps. This automation ensures that testing reports are consistent and easy to understand. Scripting with Burp Suite is made possible through its powerful and extensible API. The API exposes a wide range of functions and interfaces that allow scripts to interact with Burp Suite's features programmatically. It provides access to HTTP traffic, session management, configuration settings, and much more. The API is available in multiple programming languages, including Java, Python, and Ruby, making it accessible to a wide range of developers. To get started with scripting in Burp Suite, testers need to set up their development environment. This includes installing the Burp Extender tool, configuring the chosen programming language's environment, and setting up any required libraries or dependencies. Once the development environment is ready, testers can start writing scripts using the Burp Extender IDE, which offers features such as syntax highlighting, code completion, and debugging capabilities. Burp Suite also provides extensive documentation and examples to help testers learn how to use the API effectively. Testing and debugging scripts are essential steps in the development process. Testers should thoroughly test their scripts to ensure they behave as intended and do not introduce errors or vulnerabilities into the testing process. The Burp Extender IDE includes debugging tools that allow testers to step through their code, inspect variables, and identify issues. Scripting can be a dynamic and iterative process, with testers continually refining and expanding their scripts as they encounter new testing scenarios. In summary, automation and scripting with Burp Suite are indispensable for security professionals seeking to enhance their web application security testing capabilities. Automation can streamline tasks such as vulnerability scanning, data extraction, and reporting, *while scripting allows testers to customize and extend Burp Suite's functionality. Automation and scripting save time and ensure consistency in testing practices, making it easier to identify and remediate vulnerabilities effectively. Burp Suite's extensibility and powerful API provide the tools and resources needed to create custom scripts tailored to specific testing needs. By harnessing the power of automation and scripting, security professionals can contribute to the overall security of web applications and protect against potential threats efficiently and effectively.*

Chapter 10: Reporting and Remediation

Generating comprehensive security reports is a crucial step in the web application security testing process. These reports serve as a vital means of communicating findings, vulnerabilities, and recommendations to various stakeholders, including developers, administrators, and decision-makers. A well-structured and informative security report not only provides insights into the security posture of an application but also guides remediation efforts and helps prioritize security improvements. Next, we'll explore the key elements and best practices for generating effective and comprehensive security reports. Before delving into the details of security reporting, it's essential to understand the primary goals and objectives of these reports. Security reports aim to document the results of web application security testing, including the identification of vulnerabilities, their severity, and recommendations for remediation. These reports also serve as a record of the testing process, ensuring transparency and accountability in security assessments. When generating security reports, testers should consider the audience and tailor the content to meet their needs. For technical audiences, such as developers and system administrators, reports should provide in-depth technical details about vulnerabilities, including proof of concept (PoC) exploits and steps to reproduce. Non-technical stakeholders, on the other hand, may require a *more high-level overview of findings, potential business impacts, and recommended actions. The structure of a security report is a critical aspect that influences its effectiveness. A well-organized report typically includes the following sections: Executive Summary, Scope, Methodology, Findings, Risk Assessment, Recommendations, and Appendices. The Executive Summary is the first section of the report and provides a concise overview of the testing process and its outcomes. This section should include a brief description of the application tested, the scope of the assessment, the methodology used, and a summary of critical findings. The Executive Summary is particularly important for non-technical stakeholders who may not have the time or expertise to delve into the entire report. The Scope section defines the boundaries of the assessment and clarifies what areas of the application were included and excluded. It helps set expectations and ensures that stakeholders understand the limitations of the testing process. Methodology describes the approach and techniques used during the assessment, including tools, scripts, and manual testing procedures. Providing insight into the testing methodology adds transparency to the process and allows stakeholders to assess the rigor of the assessment. The Findings section is the heart of the security report, where identified vulnerabilities are documented in detail. Each vulnerability should be described thoroughly, including its name, description, severity, affected components, and steps to reproduce. Testers should include any relevant technical details,* such as HTTP requests and responses, to help developers understand the issue. Additionally, providing PoC exploits or code snippets can be valuable for illustrating the vulnerability's impact. The Risk Assessment section quantifies the potential business impact of each vulnerability. This assessment typically includes factors such as likelihood, impact, and exploitability. Testers may use a standardized risk scoring system, such as the Common Vulnerability Scoring System (CVSS), to assign severity ratings. The Risk Assessment helps prioritize remediation efforts, focusing on vulnerabilities with the highest risk. Recommendations provide actionable guidance on how to remediate identified vulnerabilities. These recommendations should be specific and practical, offering step-by-step instructions for mitigating the issues. Testers should also emphasize the importance of implementing security best practices to prevent similar vulnerabilities in the future. The Appendices section contains supplementary information that supports the findings and recommendations. This may include additional technical details, screenshots, logs, or any other evidence related to the assessment. Appendices provide transparency and allow stakeholders to validate the findings independently. When generating comprehensive security reports, it's essential to use clear and concise language. Avoid technical jargon or acronyms that may not be familiar to all readers. The goal is to make the report accessible and understandable to both technical and non-technical audiences. Graphics and visuals, such as charts or graphs, can enhance the report's readability and help *convey complex information more effectively. For example, a chart depicting the distribution of vulnerabilities by severity can provide a quick overview of the assessment's findings. When discussing vulnerabilities, testers should focus on the practical impact they may have on the application and the organization. This approach helps stakeholders understand the real-world consequences of the identified issues. Furthermore, providing context for vulnerabilities, such as explaining their potential exploitation scenarios, can help stakeholders assess the urgency of remediation. In addition to documenting vulnerabilities, security reports should highlight the positive aspects of the application's security posture. Acknowledging well-implemented security controls, such as strong authentication mechanisms or input validation, can provide a balanced perspective. Furthermore, security reports should emphasize the collaborative nature of the assessment. Testers should encourage open communication between security teams and development teams, fostering a cooperative approach to resolving vulnerabilities. A critical aspect of generating comprehensive security reports is ensuring that the information is accurate and well-supported. Testers should thoroughly validate their findings and avoid making assumptions or speculations. When reporting vulnerabilities, it's essential to provide concrete evidence and demonstrate the steps taken to verify the issues. This approach builds credibility and trust with stakeholders, increasing the likelihood that remediation efforts will be prioritized. When recommending remediation steps, testers should strive for clarity and specificity. Developers should be able to follow the recommendations easily and implement the necessary changes. Whenever possible, testers should provide examples or code snippets to illustrate the recommended fixes. This hands-on guidance can expedite the remediation process and reduce the likelihood of misinterpretation. Security reports should also align with industry standards and best practices. Using standardized risk scoring systems, such as CVSS, ensures consistency in severity ratings and helps stakeholders compare vulnerabilities across different assessments. Moreover, adhering to recognized reporting formats and templates can enhance the professionalism and credibility of the report. Finally, security reports should be reviewed and validated by peers or experienced professionals to ensure their accuracy and completeness. A fresh set of eyes can identify any overlooked vulnerabilities or areas where the report could be improved. In summary, generating comprehensive security reports is a critical aspect of web application security testing. These reports serve as a means of communicating findings, vulnerabilities, and recommendations to various stakeholders. Security reports should be tailored to the audience, well-structured, and clear in their language. They should provide an executive summary, document the scope and methodology, describe findings, assess risks, offer recommendations, and include supporting appendices. Security reports should use visuals to enhance readability, emphasize practical impact, acknowledge positive security measures, and promote collaboration. Accuracy and validation of findings are essential, as are clear and specific remediation recommendations aligned with industry standards. With these best practices in mind, security professionals can produce effective and actionable security reports that contribute to the improvement of web application security.* Remediation strategies and best practices are essential components of the web application security testing process. Once vulnerabilities have been identified and documented in security reports, it is crucial to address them effectively to improve the overall security posture of the application. Next, we will explore various remediation strategies and best practices that organizations can implement to mitigate security vulnerabilities and strengthen their web applications. Before delving into specific remediation strategies, it is important to establish a clear and effective process for handling vulnerabilities. This process typically involves several key steps, including vulnerability triage, prioritization, remediation, and validation. Vulnerability triage is the initial assessment of reported vulnerabilities to determine their validity and severity. This step helps filter out false positives and ensures that resources are allocated to addressing genuine security issues. Prioritization involves evaluating the severity and potential impact of vulnerabilities to determine the order in which they should be remediated. Organizations often use standardized risk scoring systems, such as the Common Vulnerability Scoring System (CVSS), to assign severity ratings and prioritize vulnerabilities accordingly.

Remediation is the process of addressing identified vulnerabilities by implementing fixes or mitigations. Validation involves verifying that the remediation efforts have successfully resolved the vulnerabilities and that no new security issues have been introduced. Establishing a well-defined vulnerability management process helps organizations efficiently address security weaknesses and maintain the security of their web applications. One of the foundational principles of effective remediation is to address the most critical vulnerabilities first. High-severity vulnerabilities, such as those that allow remote code execution or data breaches, should take precedence over lower-severity issues. By focusing on the highest-risk vulnerabilities, organizations can significantly reduce their exposure to potential attacks. Additionally, organizations should consider the ease of exploitation and potential business impact when prioritizing remediation efforts. Vulnerabilities that are easy to exploit or have a severe impact on business operations should receive immediate attention. To remediate vulnerabilities effectively, organizations should follow secure coding practices and adhere to security best practices. Secure coding involves writing code with security in mind from the outset, rather than attempting to patch security vulnerabilities after the fact. This proactive approach reduces the likelihood of introducing security issues in the first place. Secure coding practices include input validation, output encoding, proper session management, and avoiding known security pitfalls, such as SQL injection and cross-site scripting (XSS). Using security frameworks and libraries that have undergone security reviews and testing can also help organizations avoid common vulnerabilities. Incorporating security into the software development lifecycle (SDLC) is another essential aspect of effective remediation. By integrating security activities into each phase of the SDLC, organizations can identify and address security issues early in the development process. This includes performing security assessments, code reviews, and security testing as part of the development workflow. Security training and awareness programs for developers and other stakeholders can also contribute to a security-conscious culture within the organization. Implementing security controls and mechanisms is a key strategy for remediation. Organizations should employ security controls such as firewalls, intrusion detection systems, and web application firewalls (WAFs) to protect against common attack vectors. Additionally, strong authentication and access control mechanisms help prevent unauthorized access to sensitive resources. Encryption should be used to protect data in transit and at rest, reducing the risk of data breaches. Security monitoring and incident response capabilities are essential for detecting and responding to security incidents in a timely manner. Regular vulnerability assessments and penetration testing can help organizations proactively identify and address security weaknesses before they can be exploited by attackers. By conducting these assessments on a regular basis, organizations can stay ahead of emerging threats and vulnerabilities. In some cases, organizations may choose to mitigate vulnerabilities rather than fixing them immediately. Mitigation involves implementing temporary measures or compensating controls to reduce the risk associated with a vulnerability while planning and implementing a permanent fix. This approach can be useful when immediate remediation is not feasible, such as in situations where a software vendor needs time to release a patch. Organizations should carefully monitor mitigated vulnerabilities to ensure that they are addressed with permanent fixes as soon as possible. Vulnerability management and remediation should be an ongoing and iterative process. Organizations should continuously assess their web applications for new vulnerabilities and emerging threats. Regularly updating and patching software and libraries is crucial to address known vulnerabilities that may be exploited by attackers. It is also essential to monitor and respond to security alerts and incidents promptly. Furthermore, organizations should maintain an inventory of their web applications and associated assets to ensure that all potential attack surfaces are considered in the vulnerability management process. In summary, remediation strategies and best practices are essential for addressing vulnerabilities and improving the security of web applications. Organizations should establish a clear vulnerability management process that includes vulnerability triage, prioritization, remediation, and validation. Prioritizing high-severity vulnerabilities and those with a significant business impact is crucial. Secure coding practices, security training, and integration of security into the SDLC are key components *of effective remediation. Implementing security controls, encryption, and monitoring mechanisms can help protect against common attack vectors. Regular vulnerability assessments and penetration testing are essential for proactively identifying and addressing security weaknesses. Mitigation may be necessary in situations where immediate remediation is not possible. Maintaining an ongoing and iterative vulnerability management process ensures that web applications remain secure in the face of evolving threats and vulnerabilities. By following these remediation strategies and best practices, organizations can significantly reduce their exposure to security risks and protect their web applications from potential threats.*

BOOK 2
MASTERING BURP SUITE
PEN TESTING TECHNIQUES FOR WEB APPLICATIONS

ROB BOTWRIGHT

Chapter 1: Burp Suite Essentials and Setup

Installing Burp Suite on different platforms is a straightforward process that allows security professionals and penetration testers to leverage its powerful features for web application security testing. Burp Suite is a versatile and widely used tool for identifying and addressing vulnerabilities in web applications. It offers comprehensive functionality for tasks such as web crawling, vulnerability scanning, and manual testing, making it an essential tool for anyone involved in web application security. One of the first steps in getting started with Burp Suite is the installation process. Burp Suite is available for multiple platforms, including Windows, macOS, and Linux, ensuring compatibility with a wide range of operating systems. Let's explore the installation process for each of these platforms in more detail. For Windows users, the installation of Burp Suite typically involves downloading the Windows installer package from the official website. The installer is available in both 32-bit and 64-bit versions, allowing users to choose the one that matches their system architecture. After downloading the installer, users can simply double-click on it to initiate the installation process. The installer will guide users through the installation steps, including selecting the installation directory and creating shortcuts. Once the installation is complete, Burp Suite can be launched from the desktop shortcut or the Start menu. On macOS, the installation process for Burp Suite is also straightforward. Users can download the macOS installer package from the official website. The installer package is usually in the form of a disk image (DMG) file. After downloading the DMG file, users can open it by double-clicking. Inside the DMG file, there will be a Burp Suite application icon that can be dragged and dropped into the Applications folder. This action installs Burp Suite on the macOS system. Once the installation is complete, users can find Burp Suite in their Applications folder and launch it from there. For Linux users, Burp Suite provides a standalone JAR (Java Archive) file that can be executed directly from the command line. This approach offers flexibility and compatibility with various Linux distributions. To install Burp Suite on Linux, users need to download the JAR file from the official website. Once downloaded, they can open a terminal and navigate to the directory where the JAR file is located. To launch Burp Suite, users can use the following command: java -jar burpsuite_community.jar. This command starts Burp Suite on the Linux system. It's worth noting that Java Runtime Environment (JRE) or Java Development Kit (JDK) must be installed on the Linux system for this command to work. Additionally, Linux users can create desktop shortcuts or launcher scripts for Burp Suite to streamline the launch process. Installing Burp Suite on different platforms ensures that security professionals have access to its capabilities regardless of their preferred operating system. Moreover, Burp Suite offers both community and professional editions, each with its own set of features *and licensing options. The community edition is free to use and provides essential functionality for web application security testing. On the other hand, the professional edition offers advanced features, including automated scanning, reporting, and additional customization options. To access the professional edition, users need to obtain a license key from PortSwigger, the company behind Burp Suite, and enter it during the installation process or within the application itself. Once installed, Burp Suite provides a user-friendly graphical interface that simplifies the process of configuring and using its various features. Users can start by configuring proxy settings to intercept and inspect HTTP requests and responses. They can also set up options such as target scope, which defines the scope of web applications to be tested. Burp Suite allows users to define custom scope rules, ensuring that only specific parts of a web application are included in the testing process. The tool also supports various web browsers and can be configured to work as a proxy server. This enables users to direct their web traffic through Burp Suite, allowing them to intercept and analyze requests and responses. Additionally, Burp Suite provides a variety of scanning options, including automated vulnerability scanning and crawling. Users can initiate scans to identify common web application vulnerabilities such as SQL injection, cross-site scripting (XSS), and more. Burp Suite offers customizable scanning profiles, allowing users to tailor the scanning process to their specific needs. In addition to automated scanning, Burp Suite's manual testing capabilities are* highly regarded by security professionals. The tool provides an intercepting proxy that allows users to capture and modify HTTP requests and responses in real time. This feature is invaluable for identifying and exploiting vulnerabilities that may not be detected by automated scans. Burp Suite also includes a comprehensive set of tools for analyzing and manipulating web traffic. For example, the Repeater tool allows users to send repeated requests to a web application while making incremental changes to test for vulnerabilities. The Intruder tool is designed for performing automated attacks, such as brute force or parameter manipulation attacks. Furthermore, Burp Suite supports the use of extensions and plugins, which can be developed or downloaded from the Burp Extender BApp Store. Extensions can enhance Burp Suite's functionality by adding custom features and capabilities. This extensibility makes Burp Suite a versatile tool that can be adapted to meet specific testing requirements. In summary, installing Burp Suite on different platforms is a straightforward process that allows security professionals to access its powerful web application security testing capabilities. Users can choose the platform that best suits their needs, whether it's Windows, macOS, or Linux. The tool's versatility, combined with its community and professional editions, provides flexibility for various testing scenarios. Once installed, Burp Suite offers a user-friendly interface and a wide range of features, including proxying, scanning, manual testing, and extensibility through plugins. By leveraging Burp Suite, security professionals can enhance the security of web applications and identify vulnerabilities effectively. Configuring proxy settings for Burp Suite is a fundamental step in harnessing the power of this versatile web application security testing tool. Proxy settings enable Burp Suite to intercept and analyze HTTP requests and responses between a web browser and a web application, allowing security professionals to identify vulnerabilities, manipulate data, and test for potential security weaknesses. Next, we'll explore the importance of proxy settings, how to configure them in Burp Suite, and some best practices for effective proxy usage. Proxy settings serve as the core component of Burp Suite's functionality, acting as a man-in-the-middle (MITM) between the user's browser and the target web application. This interception capability is essential for conducting security assessments because it allows testers to view, modify, and analyze the traffic exchanged between the client and server. By intercepting requests and responses, security professionals gain insights into the application's behavior, revealing potential security vulnerabilities and attack vectors. To configure proxy settings in Burp Suite, users need to access the "Proxy" tab within the Burp Suite user interface. This tab provides options for configuring the proxy listener, setting up browser proxy settings, and managing various aspects of the proxy functionality. The primary configuration setting to consider is the "Proxy Listeners" section, which allows users to specify the IP address and port on which Burp Suite will listen for incoming requests. By default, Burp

Suite listens on localhost (127.0.0.1) and port 8080, but users can change these settings to suit their requirements. For example, if a user prefers to intercept traffic from a mobile device or another computer on the same network, they can configure the proxy listener to listen on the machine's external IP address. Burp Suite also supports listening on multiple ports, enabling users to simultaneously intercept traffic from different sources. To begin intercepting traffic, users need to ensure that their web browser is configured to use Burp Suite as a proxy. This configuration can typically be found in the browser's network settings or proxy settings. Users should set the browser to use the same IP address and port specified in the Burp Suite proxy listener configuration. Once the browser is configured to use Burp Suite as a proxy, all HTTP traffic generated by the browser will be directed through Burp Suite, allowing users to intercept and analyze it. Before intercepting traffic, users should consider setting up filters and rules within Burp Suite to control which requests and responses they want to intercept. For example, they can create filters to exclude static resources like images, stylesheets, and JavaScript files from interception. This reduces noise and focuses the analysis on the relevant application traffic. In addition to basic interception, Burp Suite offers various interception action options, allowing users to decide how they want to handle intercepted requests and responses. These actions include "forward," "drop," "repeat," and "modify," among others. The "forward" action allows intercepted traffic to continue to its destination without *any modification. The "drop" action discards intercepted traffic, preventing it from reaching the destination server. The "repeat" action enables users to send the intercepted request again, making it useful for testing and manipulation. The "modify" action allows users to make changes to the intercepted request before it is sent to the server. These interception actions provide flexibility and control over the testing process, enabling security professionals to simulate various attack scenarios and test application behaviors. One common use case for interception in Burp Suite is to identify and test for vulnerabilities such as cross-site scripting (XSS) and SQL injection. By intercepting and modifying parameters or payloads, testers can inject malicious code and observe the application's response. This helps in identifying whether the application is vulnerable to such attacks and whether it properly sanitizes user inputs. Additionally, security professionals can use interception to manipulate authentication mechanisms, test for session management vulnerabilities, and assess the application's resistance to various security threats. When configuring proxy settings in Burp Suite, users should also consider the importance of SSL/TLS interception. Many modern web applications use encrypted connections (HTTPS) to secure the exchange of sensitive data between clients and servers. To analyze and intercept HTTPS traffic, Burp Suite provides a feature called "SSL/TLS interception" or "HTTPS interception." This feature requires users to install Burp Suite's root CA certificate on the device or browser they are testing from. Once the certificate is installed and*

SSL/TLS interception is enabled in Burp Suite, it can intercept and decrypt encrypted traffic, making it accessible for analysis. However, it's crucial to use SSL/TLS interception responsibly and only on systems and applications that you own or have explicit permission to test. Unauthorized interception of encrypted traffic is not only unethical but may also violate legal regulations. As part of best practices, users should keep their Burp Suite installation and CA certificate up to date. PortSwigger, the company behind Burp Suite, periodically releases updates and security patches. Staying current with these updates ensures that the tool remains effective and secure. Users should also ensure that they have adequate system resources to handle the traffic being intercepted. Burp Suite can generate a substantial amount of data, especially during scans or while intercepting large responses. Insufficient system resources may lead to slowdowns or crashes. To optimize performance, users can adjust the tool's settings, such as thread and memory allocation, to match their hardware capabilities. In summary, configuring proxy settings for Burp Suite is a crucial step in conducting web application security testing. Proxy settings enable the interception and analysis of HTTP traffic, allowing security professionals to identify vulnerabilities, manipulate requests, and test application security. Users should configure proxy listeners, set up browser proxy settings, and consider filtering and interception actions to control traffic analysis effectively. SSL/TLS interception is a valuable feature for testing encrypted connections, but it should *be used responsibly and with proper authorization. Regular updates and adequate system resources are essential for maintaining the effectiveness and performance of Burp Suite. By mastering the configuration of proxy settings, security professionals can leverage Burp Suite's capabilities to improve the security of web applications and identify potential security weaknesses effectively.*

Chapter 2: Web Application Reconnaissance

Passive reconnaissance techniques are a critical component of the information-gathering phase in the field of cybersecurity. These techniques involve collecting information about a target system or network without directly interacting with it, which can provide valuable insights for security assessments. Passive reconnaissance is often the first step taken by security professionals, ethical hackers, and penetration testers when preparing for an engagement. The goal is to gather as much data as possible about the target while avoiding any actions that might trigger alarms or alerts. One of the primary sources of passive reconnaissance data is public information. This includes information that is openly available on the internet, such as the target organization's website, social media profiles, and publicly accessible documents. Security professionals can examine the target's website to gather details about the company's structure, contact information, and the technologies they use. Additionally, social media profiles of employees and key personnel may reveal valuable insights, including potential points of contact for spear-phishing attacks. Publicly accessible documents, such as job postings, presentations, or conference materials, can provide further clues about the target's technology stack and corporate culture. Search engines like Google, Bing, and Shodan can be powerful tools for passive reconnaissance. By using specific search queries, security professionals can uncover information that might not be readily available through traditional means. For example, searching for a company's subdomains, email addresses, or specific file types may yield results that expose potential vulnerabilities or areas of interest. Another passive reconnaissance technique involves analyzing Domain Name System (DNS) data. DNS is the system responsible for translating human-friendly domain names (e.g., example.com) into IP addresses (e.g., 192.168.1.1). Security professionals can perform DNS reconnaissance to gather information about the target's domain names and their associated IP addresses. Tools like DNSDumpster, Dnsmap, and others can help automate this process, providing a list of subdomains and associated IP addresses. This data can be valuable for identifying potential entry points into the target's network infrastructure. Passive reconnaissance also extends to monitoring network traffic without directly interacting with the target system. Network traffic analysis can reveal information about the target's architecture, the communication protocols in use, and potentially sensitive data being transmitted. Packet capture tools like Wireshark or network monitoring solutions can be deployed to capture and analyze network traffic. By examining packet headers and payloads, security professionals can gain insights into the target's network architecture and potential vulnerabilities. Another critical aspect of passive reconnaissance is the analysis of email information. Email addresses can be a valuable source of information, as they are often used for communication within an organization. Using tools like theHarvester or reconnaissance scripts, security professionals can discover email addresses associated with the target. These addresses can be useful for social engineering attacks, including phishing attempts. Additionally, analyzing email headers can provide clues about the target's email infrastructure, potentially revealing details about mail servers and their configurations. Passive reconnaissance efforts may also involve searching for publicly disclosed vulnerabilities related to the target's technologies. Security professionals can use vulnerability databases like the National Vulnerability Database (NVD), the Common Vulnerabilities and Exposures (CVE) database, or various exploit databases. By cross-referencing the technologies and versions in use by the target with known vulnerabilities, security professionals can identify potential weaknesses. This information can guide further testing and assessment activities. Passive reconnaissance should also encompass information about the target's internet-facing infrastructure. Tools like Nmap or Shodan can help identify open ports, services, and potentially exposed systems. Understanding the target's external attack surface is crucial for developing a comprehensive security assessment strategy. One of the key benefits of passive reconnaissance is that it is typically non-intrusive and does not raise suspicion. Unlike active reconnaissance, which involves sending probes or scans to the target, passive techniques rely on publicly available data and observations. This makes passive reconnaissance an essential initial step in the *reconnaissance phase of security assessments. While passive reconnaissance provides valuable information, it is important to note that it has limitations. It relies on publicly available data, which means it may not uncover hidden or confidential information. Additionally, the accuracy of passive reconnaissance data can vary, as it depends on the completeness and timeliness of publicly available information. Furthermore, passive reconnaissance alone does not provide a complete picture of the target's security posture. Active reconnaissance techniques, which involve interactions with the target system, are often necessary to validate findings and identify vulnerabilities that may not be evident through passive means. In summary, passive reconnaissance techniques are a vital part of the information-gathering phase in cybersecurity assessments. Security professionals use these techniques to collect valuable information about a target system or network without directly interacting with it. Passive reconnaissance leverages publicly available data sources, such as websites, search engines, DNS records, and network traffic analysis. By carefully analyzing this data, security professionals can uncover valuable insights into the target's infrastructure, potential vulnerabilities, and attack surface. While passive reconnaissance is non-intrusive and helps in the initial stages of reconnaissance, it is complemented by active reconnaissance techniques to provide a more comprehensive understanding of the target's security posture. Overall, passive reconnaissance is a crucial tool in the arsenal of security professionals and ethical* hackers, helping them gather intelligence and make informed decisions during security assessments. Active reconnaissance and enumeration are essential steps in the information-gathering phase of cybersecurity assessments. While passive reconnaissance techniques gather publicly available information without directly interacting with the target, active techniques involve actively probing the target's systems and networks to obtain more detailed and often confidential information. Active reconnaissance is the process of actively probing and scanning a target system or network to identify vulnerabilities, services, and potential points of entry. Unlike passive reconnaissance, which relies on publicly available data, active reconnaissance involves sending requests and probes to the target. One of the fundamental tools used in active reconnaissance is the port scanner. Port scanning is the process of systematically checking which network ports on a target system are open and listening for incoming connections. Port scanners send connection requests to various ports and analyze the responses to determine their status. Common port scanners like Nmap are widely used in cybersecurity assessments for active reconnaissance. By identifying open ports, security professionals can gain insights into the services running on the target system and potential vulnerabilities associated with those services. Another active reconnaissance technique is banner grabbing, which involves connecting to open ports and analyzing the information provided by the service running on that port. Banner grabbing can reveal details about the service version, software, and configuration, helping testers identify potential weaknesses. For example, if a banner grab reveals that an FTP server is running a vulnerable version, it can be a valuable piece of information for further testing. Service enumeration is closely related to active reconnaissance and involves identifying and gathering information about the services and applications running on the target system. Enumeration techniques aim to discover specific details about the services, such as users, shares, resources, and configurations. Enumeration often involves using tools like enumeration scripts, SNMP (Simple Network Management Protocol) queries, and brute force techniques to gather information. SNMP queries can provide valuable insights into network devices, their configurations, and potentially weak SNMP community strings. Brute force techniques may be employed to guess usernames and passwords, especially in scenarios where weak or default credentials are suspected. Service enumeration is critical for identifying potential vulnerabilities, misconfigurations, and weak points in the target system. Vulnerability scanning is another key aspect of active reconnaissance. Vulnerability scanners like Nessus, OpenVAS, and Qualys actively scan the target system for known vulnerabilities. These scanners leverage a vast database of vulnerability signatures to check the target's software and configurations for weaknesses. Vulnerability scanning often includes identifying outdated software versions, missing patches, and configuration errors. By detecting known vulnerabilities, security professionals can prioritize *remediation efforts and assess the risk associated with the target system. DNS enumeration is a specific form of enumeration that focuses on gathering information about the Domain Name System (DNS) infrastructure of the target organization. DNS enumeration techniques involve querying DNS servers to gather data such as hostnames, subdomains, mail servers, and their corresponding IP addresses. This information can be useful for identifying internal systems, mail servers, and potential targets for further assessment. Another technique commonly used in active reconnaissance is brute forcing. Brute force attacks involve attempting to guess usernames and passwords through trial and error. While brute force attacks can be time-consuming and resource-intensive, they can be effective when weak or default credentials are in use. Tools like Hydra and Medusa are popular choices for performing brute force attacks against various services and protocols. In addition to the technical aspects of active reconnaissance, social engineering techniques may also be employed. Social engineering involves manipulating individuals within the target organization to divulge confidential information, such as usernames, passwords, or system details. Phishing, pretexting, and tailgating are examples of social engineering tactics. These techniques can complement technical reconnaissance efforts by obtaining valuable information from employees or stakeholders. Active reconnaissance and enumeration are crucial steps in the reconnaissance phase of cybersecurity assessments because they provide a more detailed and accurate understanding of the target's security posture. However, it's essential to approach active reconnaissance with caution and permission. Unauthorized scanning or probing of systems can trigger alarms, disrupt services, and even violate laws and regulations. To conduct active reconnaissance ethically, security professionals should obtain proper authorization from the target organization or system owner. They should also adhere to established rules of engagement and follow ethical guidelines. Furthermore, active reconnaissance should be performed methodically to avoid causing harm or unintended consequences. It's essential to document findings, prioritize vulnerabilities, and communicate effectively with stakeholders to address discovered issues. In summary, active reconnaissance and enumeration are essential components of the information-gathering phase in cybersecurity assessments. Active reconnaissance techniques involve actively probing and scanning target systems to identify vulnerabilities and services. Enumeration techniques focus on gathering detailed information about services, configurations, and users. These techniques provide valuable insights into the target's security posture, helping security professionals identify potential weaknesses and prioritize remediation efforts. However, active reconnaissance should be conducted ethically, with proper authorization, and in a methodical manner to avoid unintended consequences. By mastering active reconnaissance and enumeration, security professionals can enhance their ability to assess and improve the security of target systems and networks effectively.*

Chapter 3: Identifying and Exploiting Web Vulnerabilities

Common web vulnerabilities represent a significant challenge in the field of web application security. These vulnerabilities can expose web applications to a wide range of threats, including data breaches, unauthorized access, and other malicious activities. Understanding these vulnerabilities is crucial for security professionals, developers, and anyone involved in web application development and maintenance. Next, we will provide an overview of some of the most common web vulnerabilities, shedding light on their characteristics and potential consequences. Cross-Site Scripting (XSS) is one of the most prevalent web vulnerabilities. XSS occurs when an attacker injects malicious scripts into web pages viewed by other users. These scripts can execute in the context of the victim's browser, leading to the theft of sensitive information, session hijacking, or other malicious actions. XSS vulnerabilities can be classified into three main types: Stored XSS, Reflected XSS, and DOM-based XSS. In Stored XSS, the malicious script is permanently stored on the target server, affecting all users who access the compromised page. Reflected XSS involves the attacker tricking a user into clicking a specially crafted link that contains the malicious payload. DOM-based XSS exploits vulnerabilities in the Document Object Model (DOM) of a web page to manipulate its structure and behavior. Another common web vulnerability is SQL Injection

(SQLi). SQLi occurs when an attacker injects malicious SQL queries into an application's input fields. If the application does not properly validate and sanitize user inputs, these queries can be executed, potentially revealing sensitive database information or allowing the attacker to modify the database. SQLi vulnerabilities can be particularly dangerous when they enable unauthorized access to sensitive data. Cross-Site Request Forgery (CSRF) is a vulnerability that exploits the trust that a web application has in an authenticated user's browser. In a CSRF attack, an attacker tricks a user into performing actions without their knowledge or consent while they are logged into a web application. This can lead to actions such as changing account settings, making financial transactions, or even deleting data. To prevent CSRF attacks, web applications typically use tokens to verify the authenticity of requests. Insecure Deserialization is another web vulnerability that occurs when an application improperly handles serialized data. Attackers can exploit this vulnerability to execute arbitrary code or launch denial-of-service attacks. Insecure deserialization can lead to significant security risks, especially when it enables remote code execution. Security Misconfigurations are common web vulnerabilities that arise from improperly configured web servers, databases, or application components. These misconfigurations can include default credentials, unnecessary services or features, excessive permissions, and exposed sensitive information. Attackers can leverage security misconfigurations to gain unauthorized access or escalate privileges within an application. Broken Authentication is a vulnerability that occurs when an application's authentication mechanisms are flawed or improperly implemented. Attackers can exploit this vulnerability to gain unauthorized access to user accounts, impersonate users, or manipulate session data. Broken authentication can lead to account takeover and unauthorized actions. Sensitive Data Exposure is a web vulnerability that involves the improper handling of sensitive information such as passwords, credit card numbers, or personal data. When sensitive data is not adequately protected, it can be exposed to attackers, leading to identity theft, financial fraud, and other malicious activities. To mitigate this vulnerability, web applications should use strong encryption and secure storage practices. XML External Entity (XXE) is a vulnerability that arises from improper parsing of XML input by an application. Attackers can use XXE to exploit the application's XML processing capabilities, potentially revealing sensitive information or launching denial-of-service attacks. In some cases, XXE can lead to remote code execution. Unvalidated Redirects and Forwards are vulnerabilities that occur when an application redirects or forwards a user to an untrusted or malicious site without proper validation. Attackers can use this vulnerability to trick users into visiting malicious sites, leading to phishing attacks or other malicious activities. To prevent unvalidated redirects and forwards, applications should validate and sanitize redirect URLs. These common web vulnerabilities highlight the importance of rigorous security testing and secure coding practices. Security professionals must actively scan and assess web applications for these vulnerabilities, and developers should implement secure coding techniques to prevent them. Regular security assessments, penetration testing, and code reviews can help identify and mitigate these vulnerabilities before they are exploited by malicious actors. It is essential to stay informed about emerging vulnerabilities and security best practices to protect web applications effectively. Additionally, organizations should prioritize security awareness training for their development and IT teams to ensure that everyone understands the risks and how to mitigate them. In summary, common web vulnerabilities pose significant risks to web applications and their users. Understanding these vulnerabilities, their characteristics, and potential consequences is vital for maintaining web application security. Security professionals and developers must work together to identify, address, and prevent these vulnerabilities to protect sensitive data and maintain the integrity of web applications. By implementing security best practices, conducting regular assessments, and staying informed about emerging threats, organizations can enhance the security of their web applications and reduce the risk of exploitation. SQL Injection (SQLi) is a potent and prevalent web application vulnerability that can have severe consequences when exploited by attackers. Next, we will delve into the world of SQL Injection, understanding how it works, the risks it poses, and, most importantly, how to prevent and mitigate it. At its core, SQL Injection is a technique that allows attackers to manipulate an application's SQL queries by injecting malicious SQL code. This occurs when an application fails to validate and sanitize user inputs correctly, allowing an attacker to insert SQL statements into input fields, URLs, or other data inputs. The injected SQL statements are then executed by the application's database, potentially exposing sensitive data or even allowing attackers to take control of the application or the underlying server. SQL Injection attacks can target various types of databases, including MySQL, PostgreSQL, Microsoft SQL Server, Oracle, and others. To understand SQL Injection, it's essential to grasp the concept of SQL queries. SQL (Structured Query Language) is a standard language for interacting with relational databases. Applications use SQL queries to communicate with the database and retrieve or manipulate data. Typically, an application's SQL queries are constructed dynamically by combining predefined query templates with user-supplied inputs. For example, when a user submits a search query on a website, the application might construct an SQL query like this:

sqlCopy code
SELECT * FROM products WHERE name = 'user_input';
In this query, 'user_input' represents the value provided by the user in the search field. If the application does not properly validate and sanitize user inputs, an attacker can manipulate the input in a way that alters the query's structure. Consider the following input:

```bash
' OR '1'='1
```

When inserted into the query, it transforms into:

```sql
SELECT * FROM products WHERE name = '' OR '1'='1';
```

In this altered query, the '1'='1' condition always evaluates to true, effectively bypassing any user authentication or filtering mechanisms. As a result, the attacker gains unauthorized access to data, and the SQL Injection attack is successful. SQL Injection vulnerabilities can manifest in various forms, depending on the application's logic and the developer's implementation. One common type is Union-based SQL Injection, where attackers leverage the SQL UNION operator to combine the results of a malicious query with a legitimate one. Blind SQL Injection occurs when an attacker cannot directly see the application's response but can infer information through the application's behavior. Time-based Blind SQL Injection is a variant of this, where attackers induce delays in the application's responses to extract data. Error-based SQL Injection exploits error messages generated by the database to reveal information about the database structure or data. Out-of-band SQL Injection involves sending data from the target application to an external server controlled by the attacker, allowing for data extraction. Inferential SQL Injection, also known as Blind SQL Injection, relies on the application's behavior to infer the success or failure of SQL Injection. The risks associated with SQL Injection are substantial, as successful attacks can lead to unauthorized access, data theft, data manipulation, and potentially complete compromise of the application and underlying server. Attackers can exfiltrate sensitive information, such as usernames, passwords, credit card numbers, and personal data. Furthermore, they may execute arbitrary SQL commands, potentially impacting the integrity and availability of the database and application. To prevent and mitigate SQL Injection vulnerabilities, it is crucial to adopt secure coding practices. Developers should implement input validation and sanitize user inputs effectively. Input validation involves checking user inputs to ensure they adhere to expected formats and constraints. Sanitization involves removing or escaping special characters that could be interpreted as SQL commands. Parameterized queries or prepared statements are powerful tools for preventing SQL Injection. These mechanisms separate SQL code from user inputs, making it impossible for attackers to inject malicious SQL statements directly. Web application firewalls (WAFs) are another layer of defense against SQL Injection. WAFs can detect and block SQL Injection attempts by inspecting incoming traffic and applying predefined rules. Regular security assessments and penetration testing are essential to identify and remediate SQL Injection vulnerabilities. Automated scanning tools can help discover potential weaknesses, but manual testing by experienced security professionals is often necessary for thorough assessments. Security professionals should assess not only the presence of vulnerabilities but also their potential impact and exploitability. They should verify if discovered vulnerabilities can lead to data exposure or compromise of the application. Once vulnerabilities are identified, remediation efforts should be prioritized based on their criticality. Critical vulnerabilities should be addressed immediately, while less severe ones can be addressed in subsequent releases or updates. In addition to prevention and mitigation, monitoring and logging are essential aspects of dealing with SQL Injection. Monitoring involves tracking application and database activities for suspicious behavior, while logging records important events and interactions. When an SQL Injection attempt is detected, it is crucial to log relevant details, such as the attacker's IP address, the targeted URL, and the attempted SQL statement. Logs can provide valuable information for incident response and forensic analysis. Security professionals should also keep abreast of emerging SQL Injection techniques and vulnerabilities. Attackers continually evolve their tactics, and staying informed about the latest threats is essential for effective defense. In summary, SQL Injection is a pervasive and dangerous web application vulnerability that can have severe consequences when exploited. Understanding how SQL Injection works, its various forms, and the risks it poses is critical for both security professionals and developers. Preventing SQL Injection involves adopting secure coding practices, implementing input validation, and using parameterized queries or prepared statements. Regular security assessments, monitoring, and logging are essential for *identifying and mitigating SQL Injection vulnerabilities. By taking a proactive and comprehensive approach to security, organizations can reduce the risk of SQL Injection and protect their web applications and databases from malicious exploitation.*

Chapter 4: Advanced Scanning and Crawling Techniques

Customizing web application scanning policies is a crucial aspect of conducting comprehensive security assessments. While automated scanners like Burp Suite are powerful tools for detecting vulnerabilities, they benefit greatly from fine-tuned scanning policies tailored to the specific needs and characteristics of the target application. Next, we will explore the importance of customizing scanning policies, the components that can be adjusted, and best practices for optimizing the scanning process. Automated web application scanners employ predefined scanning policies to identify vulnerabilities in web applications. These policies consist of a set of rules, configurations, and algorithms that dictate how the scanner interacts with the application, sends requests, and analyzes responses. However, no one-size-fits-all scanning policy can adequately address the diverse range of web applications and their unique architectures, technologies, and security requirements. That's where customization comes into play. Customizing scanning policies involves adjusting various parameters and configurations to align the scanner's behavior with the specific characteristics of the target application. The goal is to improve the accuracy of vulnerability detection, reduce false positives, and ensure that the scanner effectively assesses the application's security posture. One of the essential components to customize is the scope of the scan. Defining the scan scope helps the scanner focus on the relevant parts of the application while excluding areas that are not in scope. This prevents unnecessary traffic and reduces the likelihood of false positives. The scan scope can be determined by specifying the target URLs, directories, and pages that should be included or excluded from the scan. Additionally, authentication settings should be adjusted to reflect the application's authentication mechanisms. Some applications may require form-based authentication, while others rely on session cookies or other methods. Configuring authentication correctly ensures that the scanner can access authenticated parts of the application and accurately assess their security. It's important to provide valid credentials or tokens if required for authentication. During the customization process, scanning policies should also consider the application's technology stack. Different web technologies, such as PHP, ASP.NET, Java, or Ruby on Rails, may require specific configurations and payloads to identify vulnerabilities effectively. The scanner should be configured to recognize and test for vulnerabilities relevant to the technologies used by the target application. Furthermore, custom error handling settings can be adjusted to ensure that the scanner accurately interprets error responses from the application. Properly configured error handling settings help distinguish between legitimate errors and potential vulnerabilities. Customizing the crawler behavior is another critical aspect of optimizing scanning policies. The crawler is responsible for navigating the application, discovering new pages, and creating a comprehensive map of the application's structure. By adjusting crawler settings, security professionals can control the depth and breadth of the crawl. For large applications, limiting the depth of the crawl can prevent the scanner from spending excessive time on non-essential pages. Additionally, customizing the scanner's request patterns and payloads is essential for identifying vulnerabilities accurately. Different types of vulnerabilities, such as SQL Injection or Cross-Site Scripting (XSS), require specific payloads to trigger and validate. Security professionals should create and customize payloads that are relevant to the application's technologies and potential vulnerabilities. By crafting payloads that reflect the application's context, the scanner can more effectively identify vulnerabilities. Scanning policies should also account for the application's rate limiting and throttling mechanisms. Some applications may restrict the number of requests a scanner can send within a certain timeframe. Customizing scan configurations to respect these limitations helps prevent disruptions to the application and minimizes the risk of detection. Customization can also extend to fine-tuning the scanner's timing and delay settings. By adjusting the timing between requests, security professionals can emulate real-world usage patterns and reduce the likelihood of being detected as a scanning tool. This can be particularly helpful when assessing applications that employ anti-scanning measures. While customization is essential, it's important to strike a balance. Excessive customization can lead to overfitting, where the scanner becomes too specific to the target application and may miss vulnerabilities that affect similar applications. To avoid this, it's crucial to maintain a balance between customization and generality. Regularly reviewing and updating scanning policies is a best practice. Web applications evolve over time, and so do their security requirements. As new vulnerabilities emerge and technologies change, scanning policies should be adjusted accordingly. Security professionals should stay informed about the latest threats, vulnerabilities, and security best practices to keep scanning policies up to date. Additionally, continuous monitoring of scan results and false positives can provide valuable insights into the effectiveness of scanning policies. When false positives occur, it's an opportunity to refine and improve the policies to reduce such occurrences in the future. Finally, collaboration between security professionals, developers, and application owners is essential for effective policy customization. Developers can provide insights into the application's architecture and technologies, helping security professionals tailor scanning policies more accurately. Application owners can provide guidance on the critical parts of the application and any specific security concerns. In summary, customizing web application scanning policies is a vital step in conducting effective security assessments. By adjusting scanning parameters, configurations, and payloads, security professionals can optimize the scanner's performance and accuracy. Customization helps ensure that the scanner aligns with the specific characteristics and security requirements of the target application. However, customization should be done judiciously to strike a balance between specificity and generality. Regular updates and collaboration with stakeholders are key to maintaining effective scanning policies that adapt to evolving security needs and technologies. By following best practices and staying informed about the latest threats, security professionals can enhance the security of web applications and reduce the risk of vulnerabilities going undetected. Handling complex web application structures is a challenge that security professionals often encounter in their efforts to assess and protect web applications. Complexity can arise from a variety of factors, including the application's size, architecture, technology stack, and the integration of multiple components and services. Next, we will explore strategies and best practices for effectively handling and securing complex web application structures. Complex web application structures can be daunting, but breaking them down into manageable components is the first step toward effective assessment and protection. One common approach is to create an application map or inventory that identifies all the components, modules, and interfaces within the application. This map helps security professionals gain a comprehensive understanding of the application's structure and dependencies. Furthermore, understanding the application's architecture and technology stack is crucial. Different technologies, such as microservices, APIs, single-page applications (SPAs), and serverless architectures, present unique challenges and require tailored security assessments. By gaining insights into the application's architecture, security professionals can prioritize their efforts and focus on critical areas. For large and complex web applications, it's essential to develop a testing strategy that accounts for different aspects of the application. This may involve a combination of automated scanning, manual testing, and specialized tools. Automated scanning tools like Burp Suite and OWASP ZAP can help identify common vulnerabilities across a wide range of pages and components. Manual testing, on the other hand, allows security professionals to dig deeper into specific areas, test custom functionalities, and explore potential attack vectors. Incorporating specialized tools for areas like API testing, mobile application testing, and cloud security assessment is also essential when dealing with complex structures. In addition to testing, continuous monitoring plays a crucial role in handling complex web application structures. Complex applications are more prone to vulnerabilities that may arise from frequent updates, changes in configuration, or integration with third-party services. Monitoring helps security professionals stay vigilant and detect anomalies or security incidents in real-time. Another challenge posed by complex web application structures is the need for effective collaboration between security professionals and other stakeholders, including developers, system administrators, and application owners. Clear communication and collaboration are essential to ensure that security assessments are aligned with the application's goals and requirements. Developers can *provide valuable insights into the application's inner workings and offer guidance on mitigating vulnerabilities. System administrators may help configure security controls and manage the application's infrastructure. Application owners can provide context on the application's critical functions and user data, helping security professionals prioritize their efforts. Understanding the data flow within a complex web application is paramount to its security. Security professionals should identify the paths that data takes through the application, from input to storage and retrieval. This involves mapping how data is processed, transmitted, and stored across different components and services. By tracing the data flow, security professionals can pinpoint potential points of exposure or data leakage. Additionally, it's crucial to assess the security of data storage and encryption mechanisms, especially when sensitive information is involved. Dealing with third-party integrations is a common aspect of complex web application structures. External components, such as payment gateways, social media APIs, or cloud services, introduce potential security risks. Security professionals should assess the security posture of these third-party integrations, verify that they are configured securely, and monitor for any changes or vulnerabilities. API security is a significant consideration in complex web application structures. With the proliferation of RESTful APIs and GraphQL, understanding how data is exchanged between the front-end and back-end components is essential. Security assessments should include API testing to identify vulnerabilities like insecure API endpoints, authentication issues, and improper data validation. Moreover, securing APIs may require the implementation of access controls, rate limiting, and thorough input validation. Complex web application structures often involve the integration of multiple microservices or components. Microservices architecture can offer scalability and flexibility, but it also introduces new challenges in terms of security. Each microservice should be assessed individually for vulnerabilities, and the interactions between microservices must be scrutinized to prevent unauthorized access or data leakage. Serverless architectures, which rely on cloud-based functions, present their own security considerations. Security professionals should evaluate serverless functions for potential security weaknesses and assess how they interact with other components. In addition to assessing individual components, holistic security assessments of the entire complex web application are crucial. This involves simulating real-world attack scenarios and considering the impact of vulnerabilities in interconnected components. By thinking like an attacker and identifying potential attack vectors, security professionals can uncover vulnerabilities that may not be apparent in isolated assessments. In summary, handling complex web application structures requires a strategic and multifaceted approach to security. Security professionals should create an application map, understand the architecture and technology stack, and develop a testing strategy that combines automated scanning, manual testing, and specialized tools. Continuous monitoring, collaboration with stakeholders, and data flow analysis are essential aspects of securing complex applications. Third-party integrations, API security, microservices, and serverless architectures should all be evaluated for potential vulnerabilities. A holistic approach to security assessments that considers the entire application's attack surface is critical in uncovering and mitigating vulnerabilities in complex web application structures. By following best practices and staying informed about emerging threats, security professionals can effectively handle the challenges posed by complex web applications and protect them from potential security risks.*

Chapter 5: Intercepting and Analyzing Web Traffic

Intercepting HTTP requests and responses is a fundamental skill for web application security professionals. It allows you to inspect and manipulate the data exchanged between a client (usually a web browser) and a web server. Next, we will explore the importance of intercepting HTTP traffic, the tools and techniques involved, and the scenarios where this skill is invaluable. When you access a web application, your web browser sends HTTP requests to the server, which responds with HTTP responses containing the requested web pages and resources. Interception comes into play by placing an intermediary, often called a proxy, between your browser and the server. This proxy acts as a middleman, intercepting all HTTP traffic and allowing you to view, modify, and even block the requests and responses. Interception serves several critical purposes in the realm of web application security. Firstly, it enables you to examine the content and structure of HTTP traffic. By inspecting the requests and responses, you gain insights into how the application communicates with the server, what data is exchanged, and what security mechanisms are in place. Understanding the underlying HTTP traffic is essential for identifying vulnerabilities and security issues. Secondly, interception allows you to identify security flaws and vulnerabilities in real-time. You can scrutinize HTTP requests for potential issues such as insecure parameters, missing input validation, or suspicious payloads. Similarly, you can analyze HTTP responses for signs of vulnerabilities like Cross-Site Scripting (XSS) or sensitive data exposure. This real-time analysis empowers security professionals to detect and address security problems as they occur. Thirdly, interception facilitates the testing of security controls and mechanisms. You can test how the application handles various security features, such as authentication, session management, and access controls, by intercepting and modifying requests. This helps ensure that the security controls are working correctly and effectively protecting the application. Interception also plays a crucial role in security assessments and penetration testing. By intercepting and manipulating HTTP traffic, security professionals can simulate various attack scenarios and assess the application's resilience to common web vulnerabilities. For instance, you can inject malicious payloads, tamper with cookies, or test for SQL Injection by modifying requests and observing how the application responds. In addition to identifying vulnerabilities, interception assists in verifying the effectiveness of security mitigations. You can validate whether security patches or configurations have been correctly implemented by examining the responses and behavior of the application. Moreover, interception is a valuable tool for debugging and troubleshooting. When unexpected behavior or errors occur in a web application, intercepting the traffic allows you to examine the requests and responses to pinpoint the root cause. This can significantly expedite the process of diagnosing and fixing issues. To intercept HTTP traffic effectively, you need a proxy tool specifically designed for this purpose. One of the most popular tools in this category is Burp Suite, which provides a dedicated proxy component. Burp Suite acts as an intercepting proxy, allowing you to capture and manipulate HTTP requests and responses. To use Burp Suite or a similar tool, you need to configure your web browser to route its traffic through the proxy. Once set up, the proxy captures all HTTP traffic between the browser and the web server. You can then view the intercepted requests and responses in the proxy tool's user interface. Interception often occurs in a proxy's intercept mode, where it pauses traffic, and the user can decide whether to allow, modify, or block each request. For example, when a request is intercepted, you can inspect its content, headers, and parameters. You can also modify any part of the request, such as changing parameters or adding custom headers. Once you've made your modifications, you can choose to forward the request to the server or drop it altogether. Interception tools like Burp Suite offer additional features to enhance your workflow. For instance, you can save intercepted requests and responses for later analysis, create and manage custom rules for automated traffic manipulation, and set breakpoints to intercept requests based on specific conditions. Interception is not limited to capturing traffic between your browser and a web server. It can also be applied to other components within a web application ecosystem, such as APIs, mobile applications, and IoT devices. For API testing, you can intercept requests sent from a mobile application or a scripting tool and analyze the data being exchanged. This is particularly useful when assessing the security of mobile applications or RESTful APIs. Intercepting requests and responses from IoT devices can help identify vulnerabilities in their communication protocols and data handling mechanisms. Moreover, interception can be valuable in monitoring and securing cloud-based applications and services. You can intercept requests sent to cloud providers, inspect data transfers, and ensure that sensitive information is transmitted securely. However, it's crucial to use interception responsibly and within legal and ethical boundaries. Unauthorized interception of someone else's traffic is a violation of privacy and may be illegal. In the context of web application security assessments and penetration testing, you should always obtain proper authorization and permissions before intercepting traffic. Furthermore, interception should be used as a means to improve the security of web applications and not for malicious purposes. In summary, intercepting HTTP requests and responses is a vital skill for web application security professionals. It enables you to inspect, manipulate, and assess the security of web applications in real-time. Interception tools like Burp Suite serve as powerful aids for capturing and analyzing HTTP traffic. By understanding the importance of interception and employing it responsibly, security professionals can effectively identify vulnerabilities, test security controls, and enhance the security of web applications and their components. Analyzing web traffic is a critical aspect of web application security assessments, and Burp Suite offers a wide range of tools and features to help security professionals in this endeavor. Next, we will delve into the various Burp Suite tools designed for analyzing web traffic and understanding how they can be used effectively. One of the core functionalities of Burp Suite is its proxy tool, which allows you to intercept and analyze HTTP requests and responses between the client (typically a web browser) and the web server. To initiate interception, you need to configure your web browser to route its traffic through Burp Suite's proxy. Once the interception is set up, Burp Suite captures all HTTP traffic, enabling you to inspect, modify, and analyze it. The Proxy Intercept tool within Burp Suite allows you to pause traffic, review each request and response, and decide whether to allow, modify, or block them. This granular control is invaluable for security assessments as it enables you to scrutinize each interaction and assess potential security vulnerabilities. When intercepting traffic, you can inspect various components of HTTP requests and responses. This includes the request method (GET, POST, etc.), headers (such as User-Agent and Referer), parameters (such as form fields and query strings), and the response status code and body. By examining these elements, you gain insights into how the application communicates with the server, what data is exchanged, and whether there are any security concerns. Additionally, you can use Burp Suite's Intercept tool to manipulate requests and responses. For instance, you can modify parameter values, change headers, or insert custom payloads to test for vulnerabilities like Cross-Site Scripting (XSS) or SQL Injection. This ability to manipulate traffic in real-time is invaluable for assessing the application's security controls and identifying potential issues. Another powerful feature of Burp Suite is its Repeater tool, which allows you to resend and modify individual requests. This tool is particularly useful when you want to test specific functionalities or endpoints repeatedly. You can take a captured request from the proxy history, send it to the Repeater tool, and then make adjustments as needed before resending it to the server.

This saves time and effort when conducting repetitive tests or exploring different attack vectors. Burp Suite also provides the Intruder tool, which automates the process of sending a large number of requests with variations to test for vulnerabilities. This tool is incredibly versatile and can be used for a wide range of security testing tasks. For instance, you can use it to perform brute-force attacks, fuzz parameter values, or test for SQL Injection by injecting payloads into parameters. The Intruder tool allows you to define attack payloads, positions within the request, and payload processing rules, giving you fine-grained control over the testing process. In addition to the Proxy, Repeater, and Intruder tools, Burp Suite offers the Sequencer tool for analyzing the randomness and quality of tokens or session identifiers. This tool helps identify weak or predictable tokens that could be exploited by attackers. By analyzing a set of tokens, Burp Suite calculates their entropy and provides insights into their unpredictability. Furthermore, Burp Suite provides a comprehensive scanning feature that automates the detection of web vulnerabilities. The Scanner tool analyzes the entire application by sending various payloads and observing the application's responses. It identifies vulnerabilities such as SQL Injection, Cross-Site Scripting (XSS), and many others. While the Scanner tool is automated and can uncover a wide range of vulnerabilities, it's essential to supplement it with manual testing and validation to reduce false positives and uncover subtle issues. Burp Suite's extensibility is another valuable aspect when it comes to analyzing web traffic. The tool supports extensions that can be developed to add custom functionality. Security professionals can create *extensions to automate repetitive tasks, customize scanning, or integrate with other security tools. This extensibility allows you to tailor Burp Suite to your specific needs and streamline your workflow. In summary, analyzing web traffic with Burp Suite tools is a fundamental part of web application security assessments. Burp Suite's proxy, intercept, repeater, intruder, sequencer, and scanning capabilities provide a comprehensive suite of tools for inspecting and testing web applications. These tools enable you to understand how the application communicates with the server, identify vulnerabilities, and assess the effectiveness of security controls. When conducting security assessments, it's essential to use a combination of automated scanning and manual testing to achieve the best results. By leveraging Burp Suite's features and extensibility, security professionals can efficiently analyze web traffic and uncover vulnerabilities, ultimately enhancing the security of web applications.*

Chapter 6: Web Application Authentication and Authorization

Testing authentication mechanisms is a crucial step in assessing the security of a web application. Authentication is the process of verifying the identity of users or entities trying to access a system, and it plays a fundamental role in protecting sensitive data and functionality. Next, we will explore the various aspects of testing authentication mechanisms, including common vulnerabilities, testing strategies, and best practices. Authentication mechanisms are designed to ensure that only authorized users gain access to an application or system. These mechanisms typically involve the use of credentials, such as usernames and passwords, to validate a user's identity. However, authentication can take many forms, including multi-factor authentication (MFA), single sign-on (SSO), biometric authentication, and more. The effectiveness of authentication mechanisms is paramount to the overall security of a web application. Weak or improperly implemented authentication can lead to unauthorized access, data breaches, and various security risks. One of the primary goals of testing authentication mechanisms is to identify vulnerabilities that could be exploited by attackers to bypass or subvert the authentication process. One common vulnerability is weak password policies, which may allow attackers to guess or brute-force user passwords. Testing for weak or default *passwords, password reuse, and the absence of password complexity requirements is essential. Another vulnerability is improper session management, which can lead to session fixation or session hijacking attacks. Testing for session-related vulnerabilities involves examining how sessions are created, managed, and destroyed and identifying weaknesses in session tokens or cookies. Cross-Site Request Forgery (CSRF) is a potential threat to authentication mechanisms, as attackers can trick users into making unintended actions. Testing for CSRF vulnerabilities involves sending forged requests and checking if the application enforces anti-CSRF tokens or other protective measures. Inadequate account lockout mechanisms may expose the application to brute-force attacks. Testing account lockout policies, such as the number of failed login attempts allowed and the lockout duration, helps assess the application's resistance to brute-force attacks. Testing authentication mechanisms often includes assessing the implementation of multi-factor authentication (MFA) if it is in use. MFA adds an extra layer of security by requiring users to provide multiple forms of identification, such as a password and a one-time code sent to their mobile device. Testing MFA involves verifying that it is correctly implemented and cannot be bypassed. Single sign-on (SSO) is another authentication mechanism that allows users to access multiple applications with a single set of credentials. Testing SSO implementations ensures that users' authentication tokens cannot be easily forged or stolen to gain unauthorized access to connected applications.*

Biometric authentication, such as fingerprint or facial recognition, is becoming increasingly common in mobile applications. Testing biometric authentication mechanisms involves examining how biometric data is stored, transmitted, and verified to prevent unauthorized access. When testing authentication mechanisms, it's essential to follow a systematic approach. One common strategy is to start with the most straightforward and commonly used authentication mechanisms and gradually move to more complex ones. Testing begins with basic username and password authentication, where security professionals attempt to identify weak passwords, password policy issues, and other common vulnerabilities. Once the basic authentication is thoroughly tested, security professionals can move on to testing more advanced mechanisms like MFA, SSO, and biometric authentication. Throughout the testing process, it's crucial to document findings, including identified vulnerabilities, their severity, and potential impact. Security professionals should also consider the role of authorization in the context of authentication testing. Authorization determines what authenticated users are allowed to do within an application. While authentication verifies a user's identity, authorization defines their level of access and permissions. Testing authorization mechanisms involves examining how user roles, permissions, and access controls are enforced. It is essential to ensure that authenticated users cannot perform actions or access resources they are not authorized to. Beyond testing, best practices for securing authentication mechanisms should be followed during application development. These practices include implementing strong password policies, enforcing session management best practices, and using secure protocols for transmitting credentials and tokens. Implementing multi-factor authentication and regularly reviewing and updating authentication mechanisms are also recommended. Furthermore, conducting periodic security assessments, such as penetration testing and vulnerability scanning, can help identify and address authentication vulnerabilities. In summary, testing authentication mechanisms is a critical component of web application security assessments. Weaknesses in authentication can lead to unauthorized access and security breaches, making it essential to identify and address vulnerabilities. Testing should cover common authentication vulnerabilities, such as weak passwords, session management issues, and CSRF vulnerabilities. Security professionals should follow a systematic approach, document findings, and consider the role of authorization in the testing process. Best practices for securing authentication mechanisms should also be implemented to enhance the overall security of the web application. By thoroughly testing and securing authentication mechanisms, organizations can mitigate the risk of unauthorized access and protect sensitive data and resources. Bypassing authentication and authorization controls is a critical aspect of web application security testing. Authentication controls are designed to verify the identity of users or entities attempting to access an *application, while authorization controls determine what actions or resources authenticated users are allowed to access. In some cases, security professionals need to assess the effectiveness of these controls by attempting to bypass them. Understanding how attackers might exploit vulnerabilities to gain unauthorized access is essential for securing web applications. Authentication and authorization bypasses can have severe consequences, as they can lead to unauthorized data access, account takeovers, and potential exposure of sensitive information. Testing the robustness of authentication and authorization controls helps identify and remediate vulnerabilities before malicious actors can exploit them. One common method of bypassing authentication controls is through credential-based attacks. These attacks involve attempting to log in with stolen or guessed credentials. For instance, attackers may try commonly used passwords, perform brute-force attacks, or use credential stuffing techniques to exploit reused passwords. Security professionals need to assess the application's resistance to these attacks by attempting to log in with invalid credentials and observing how the application responds. Weak authentication mechanisms, such as those that do not enforce password complexity or account lockout policies, are particularly vulnerable. Another technique for bypassing authentication controls is session fixation. In this attack, an attacker sets a user's session identifier to a known value, potentially compromising the user's session. Testing for session fixation involves examining how session identifiers are* generated, managed, and validated. Security professionals should assess whether it is possible to set a session identifier and hijack a user's session. In some cases, authorization controls may be vulnerable to bypasses that allow unauthorized access to restricted resources or actions. One such bypass is the direct object reference (DOR) attack. In a DOR attack, an attacker manipulates input parameters, such as URLs or form fields, to gain access to resources or perform actions not intended for their role. Security professionals need to assess whether the application correctly enforces access controls and does not rely solely on client-side security measures. Another authorization bypass technique is privilege escalation. In this scenario, an attacker with limited access attempts to elevate their privileges to gain more extensive permissions or access to sensitive areas of the application. Testing for privilege escalation involves exploring the application's role-based access controls and assessing whether there are vulnerabilities that allow a user to gain unauthorized privileges. Session-based attacks, such as session fixation or session hijacking, can also lead to unauthorized access by exploiting vulnerabilities in session management. Security professionals should assess whether the application correctly invalidates or reassigns session identifiers after authentication or authorization changes. It's crucial to understand that the techniques mentioned here should only be performed with proper authorization and within a controlled testing environment. Unauthorized testing of authentication and authorization controls can lead to security incidents and legal consequences. Additionally, it's essential to consider ethical hacking principles and responsible disclosure when conducting these assessments. Testing authentication and authorization bypasses involves a systematic approach. Security professionals need to identify the attack surface, including login forms, session management mechanisms, and access control components. Once the attack surface is defined, they can employ various testing techniques to assess the security of these controls. Credential-based attacks should include password guessing, brute-force attacks, and password spraying to assess password policies and account lockout mechanisms. Testing for session fixation should involve manipulating session identifiers and observing how the application responds. Direct object reference attacks can be tested by manipulating input parameters, while privilege escalation testing may involve attempting to access restricted resources. In addition to manual testing, security professionals can leverage automated tools to identify common authentication and authorization vulnerabilities. Such tools can help identify misconfigurations, weak password policies, and known vulnerabilities. However, manual testing should complement automated scans to uncover more subtle issues. Furthermore, testing should consider different user roles and scenarios to ensure that all potential vulnerabilities are addressed. In summary, bypassing authentication and authorization controls is a vital aspect of web application security testing. Understanding how attackers might exploit vulnerabilities in these controls helps identify and *remediate security issues before they can be exploited maliciously. Testing should be conducted systematically, encompassing various techniques and scenarios while adhering to ethical and legal standards. By thoroughly assessing the robustness of authentication and authorization controls, organizations can enhance the security of their web applications and protect against unauthorized access and data breaches.*

Chapter 7: Attacking Client-Side Security

Cross-Site Scripting (XSS) attacks are a prevalent and potentially severe web application security vulnerability. These attacks occur when an attacker injects malicious scripts into web pages viewed by other users. The injected scripts can execute in the context of the victim's browser, potentially compromising their session or stealing sensitive information. Understanding XSS attacks is essential for both security professionals and developers to protect web applications effectively. XSS attacks typically target the input fields of web applications, such as search bars, contact forms, or comment sections. Attackers inject malicious code, often written in JavaScript, into these input fields, which is then stored or reflected back to other users. Stored XSS attacks occur when the injected script is permanently stored on the server and executed whenever a user visits the affected page. Reflected XSS attacks involve the script being immediately executed when a user clicks on a malicious link or visits a compromised page. The impact of XSS attacks can range from defacing web pages and stealing user cookies to performing actions on behalf of users without their consent. One common type of XSS attack is called "alert-based" or "pop-up" XSS. In this scenario, an attacker injects a script that triggers an alert or pop-up message when a user visits a page containing the malicious script. While this type of attack may seem harmless, it serves as a proof of concept for the attacker, demonstrating their ability to execute arbitrary code in the user's browser. Another form of XSS attack is "persistent" or "stored" XSS. In persistent XSS, the attacker injects a malicious script that is permanently stored on the server and displayed to other users when they visit the affected page. This can lead to the compromise of user accounts, data theft, or the spreading of malware to unsuspecting visitors. Reflective XSS attacks, on the other hand, involve the injection of a script that is immediately executed when a user clicks on a specially crafted link or visits a manipulated page. These attacks are often used in phishing campaigns to steal user credentials or session cookies. To mitigate XSS attacks, web developers must employ secure coding practices. One fundamental preventive measure is input validation and sanitization. Input validation ensures that data entered by users adheres to expected formats and prevents malicious code from being accepted. Sanitization involves cleansing input data by removing or escaping potentially harmful characters, such as angle brackets and quotes. Developers should also implement output encoding, which ensures that any data displayed on a web page is properly encoded to prevent the execution of injected scripts. Using security libraries and frameworks that include built-in protections against XSS can simplify this process. Additionally, setting the "HttpOnly" flag on session cookies prevents JavaScript from accessing them, reducing the risk of session hijacking. Content Security Policy (CSP) headers provide another layer of defense by allowing web developers to specify which scripts are allowed to run on a page. CSP headers can restrict the execution of scripts to trusted sources, mitigating the impact of XSS attacks. Regularly updating web application software and libraries is essential, as developers often release security patches to address known vulnerabilities. To detect and prevent XSS attacks, web application security testing should be conducted regularly. Security professionals can use automated tools to scan web applications for potential vulnerabilities. Manual testing, such as attempting to inject scripts into input fields, is also crucial to identify and remediate XSS vulnerabilities. Security researchers and ethical hackers often use "payloads" to test for XSS vulnerabilities, attempting to inject scripts that simulate malicious behavior. Web application firewalls (WAFs) can also help mitigate XSS attacks by monitoring and filtering incoming traffic for suspicious patterns or payloads. Education and awareness are key in the fight against XSS attacks. Both developers and users should be educated about the risks and consequences of XSS vulnerabilities. Developers should receive training on secure coding practices, while users should be cautious about clicking on suspicious links or sharing sensitive information on untrusted websites. In summary, Cross-Site Scripting (XSS) attacks represent a significant threat to web applications and their users. Understanding the different types of XSS attacks, their potential impact, and how to prevent them is crucial for both developers and security professionals. By implementing secure coding practices, conducting regular security testing, and raising awareness, organizations can effectively *defend against XSS attacks and protect their web applications and users from harm. Cross-Site Request Forgery (CSRF) attacks pose a significant threat to the security of web applications. These attacks occur when an attacker tricks a user into unknowingly making an unwanted request to a web application on which the user is authenticated. The victim's browser sends the request, including any associated session cookies, making it appear as if it originated from the user. CSRF attacks can have severe consequences, such as changing a user's password, transferring funds without authorization, or performing actions on behalf of the victim. Understanding how CSRF attacks work and how to defend against them is crucial for web developers and security professionals. The fundamental principle behind CSRF attacks is exploiting the trust that web applications have in the user's browser. Web applications often rely on session cookies to identify and authenticate users, assuming that the browser will only send requests on behalf of the authenticated user. However, this trust can be exploited by attackers who trick users into executing malicious actions without their knowledge or consent. For example, consider an online banking application that allows users to transfer funds by clicking a link. An attacker can craft a malicious website containing an invisible form that submits a fund transfer request to the banking application when the victim visits the site. If the victim is already authenticated in the banking application, their browser will send the request with their session cookie, making it appear as if the user initiated the transfer. To defend against CSRF attacks,* web developers should implement anti-CSRF measures. One common approach is to include a unique and random token with each request that modifies the application's state. This token, known as a CSRF token, is tied to the user's session and must be presented with the request for it to be considered valid. If the token is missing or incorrect, the request is rejected. By requiring the presence of a CSRF token, attackers cannot forge requests since they would not have access to the user's session-specific token. Additionally, developers should ensure that their web applications do not perform sensitive actions based solely on GET requests, which can be more easily exploited in CSRF attacks. Instead, actions that modify the application's state should use POST requests, which are less likely to be triggered inadvertently by a user. Another recommended defense is to use the "SameSite" attribute for cookies. The "SameSite" attribute restricts when cookies are sent by the browser, reducing the risk of CSRF attacks. Developers can set cookies as "SameSite=Strict" to prevent them from being sent in cross-site requests, or as "SameSite=Lax" to allow some cross-origin requests. It's important to note that the effectiveness of these countermeasures depends on their proper implementation. CSRF tokens should be generated securely, tied to the user's session, and checked rigorously for each request. Developers should also avoid using predictable or easily guessable tokens. Testing web applications for CSRF vulnerabilities is crucial. Security professionals can simulate CSRF attacks by crafting malicious web pages or scripts that attempt to perform unauthorized actions on the target application. These tests help identify whether the application enforces anti-CSRF measures effectively. Regular security assessments, such as penetration testing and code reviews, can uncover and address CSRF vulnerabilities. Additionally, web application firewalls (WAFs) can provide an extra layer of defense by detecting and blocking suspicious cross-site requests. Education plays a vital role in preventing CSRF attacks. Both developers and users should be aware of the risks associated with CSRF and the best practices for mitigating them. Developers should receive training on secure coding practices, emphasizing the importance of CSRF protection. Users should be cautious when clicking on links or interacting with web applications, especially if they are not logged out or have active sessions. In summary, Cross-Site Request Forgery (CSRF) attacks exploit the trust that web applications have in the user's browser to trick users into executing malicious actions. Understanding how CSRF attacks work and implementing effective defenses, such as CSRF tokens and secure cookie settings, is essential for securing web applications. Testing for CSRF vulnerabilities and raising awareness among developers and users are key steps in preventing CSRF attacks and protecting the integrity of web applications.

Chapter 8: Exploiting API and Web Services

Identifying and testing API endpoints is a critical aspect of web application security assessment. APIs, or Application Programming Interfaces, enable communication and data exchange between different software components. Modern web applications often rely heavily on APIs to fetch and send data, making them an attractive target for attackers. By identifying and testing API endpoints, security professionals can assess the security of the entire web application ecosystem. API endpoints are specific URLs or routes that allow clients, such as web browsers or mobile apps, to interact with a web service or application. These endpoints define the functionality and data that can be accessed and manipulated. To identify API endpoints, security professionals typically use a combination of manual and automated techniques. Manual exploration involves reviewing the application's documentation, source code, and network traffic to identify potential endpoints. Automated tools can also help discover hidden or undocumented endpoints by scanning the application for known patterns and behaviors. When testing API endpoints, security professionals aim to identify vulnerabilities and security weaknesses that could be exploited by attackers. One common vulnerability in API endpoints is improper authentication and authorization. It's crucial to ensure that only authorized users or systems can access sensitive API endpoints and that they have the appropriate *permissions. Testing for authentication issues involves attempting to access protected endpoints without proper credentials and observing how the application responds. Another critical aspect of API security is input validation and data validation. API endpoints should validate and sanitize input data to prevent common security threats such as SQL injection, cross-site scripting (XSS), and code injection. Security professionals should test for input validation flaws by submitting malicious data or payloads to the API and evaluating the application's response. API endpoints should also be protected against rate limiting and brute-force attacks. Rate limiting controls the number of requests a user or client can make to an endpoint within a specific time frame, reducing the risk of abuse. Testing for rate limiting issues involves sending a large number of requests to an endpoint and observing whether the application enforces rate limits. Additionally, API endpoints should implement proper error handling and not leak sensitive information in error responses. Security professionals should assess how the application handles unexpected or malformed requests and whether it discloses unnecessary details that could aid attackers. Testing for error handling vulnerabilities involves submitting malformed or malicious requests to the API and examining the error responses. Data privacy and protection are critical considerations when testing API endpoints. Security professionals should ensure that sensitive data, such as user credentials and personal information, is transmitted and stored securely. Testing for data privacy involves intercepting and analyzing API traffic to identify potential data leakage or exposure. Another important aspect of API security is session management. APIs that involve user authentication should manage sessions securely, using techniques such as session tokens or JSON Web Tokens (JWTs). Security professionals should assess the implementation of session management in API endpoints to ensure that it is not susceptible to session fixation, session hijacking, or session timeout issues. When testing API endpoints, it's essential to consider both the functional and non-functional aspects of security. Functional security testing focuses on identifying vulnerabilities and weaknesses in the API's logic and functionality. Non-functional security testing assesses aspects such as performance, availability, and scalability under security-related conditions. Load testing and penetration testing are examples of non-functional security testing techniques that can be applied to API endpoints. Load testing evaluates how well the API handles a high volume of requests, simulating traffic spikes or DDoS attacks. Penetration testing involves attempting to exploit vulnerabilities and gain unauthorized access to the API. To ensure comprehensive API endpoint testing, security professionals should follow a structured approach. They should create a testing plan that outlines the scope, objectives, and testing methodology. The plan should specify the tools and techniques to be used, as well as the expected outcomes. During testing, security professionals should document their findings, including any vulnerabilities, weaknesses, or misconfigurations discovered. After identifying and testing API endpoints,* security professionals should collaborate with developers and application owners to remediate any issues. This collaboration is essential for ensuring that security vulnerabilities are addressed promptly and that the application is protected effectively. In summary, identifying and testing API endpoints is a crucial component of web application security assessment. APIs play a central role in modern web applications, and their security must be thoroughly evaluated. By following a structured testing approach and considering both functional and non-functional security aspects, security professionals can help protect web applications from a wide range of threats and vulnerabilities. API security is of paramount importance in today's digital landscape, where web applications rely heavily on APIs to interact with each other and share data seamlessly. APIs, or Application Programming Interfaces, serve as bridges between different software systems, enabling them to communicate, exchange information, and execute functions. However, the increased reliance on APIs also makes them a prime target for cyberattacks, emphasizing the need for robust API security practices. Next, we will explore API security best practices that are essential for safeguarding your applications and data. First and foremost, authentication is a cornerstone of API security. Proper authentication ensures that only authorized users or systems can access your APIs. One of the most common authentication methods for APIs is token-based authentication. When a client requests access to an API, it must provide a unique token, typically in the form of an API key or a JSON Web Token

(JWT), to prove its identity. The API server then verifies the token's validity before processing the request. This adds a layer of security by ensuring that only clients with valid tokens can interact with the API. Additionally, implementing strong password policies and enforcing multi-factor authentication (MFA) for API access can further enhance security. Authorization is closely tied to authentication and plays a crucial role in API security. Once a client is authenticated, authorization determines what actions and resources it can access. Role-based access control (RBAC) is a widely used authorization model that assigns specific roles or permissions to users or client applications. By assigning appropriate roles and permissions, you can restrict access to sensitive API endpoints and data, reducing the risk of unauthorized access. Least privilege principle should be applied when defining roles and permissions, ensuring that clients have access only to the resources necessary for their intended functions. API security also entails protecting against common web application vulnerabilities, such as SQL injection, Cross-Site Scripting (XSS), and Cross-Site Request Forgery (CSRF). Input validation and data validation are crucial to prevent these vulnerabilities. API endpoints should validate and sanitize input data to ensure it adheres to expected formats and does not contain malicious code. Furthermore, output encoding should be applied to prevent the execution of injected scripts in API responses. Security testing, including vulnerability scanning and penetration testing, should be performed regularly to identify and remediate vulnerabilities in your APIs. When it comes to secure communication, the use of HTTPS (Hypertext Transfer Protocol Secure) is non-negotiable. HTTPS encrypts data transmitted between the client and the API server, preventing eavesdropping and data interception. API endpoints should enforce the use of HTTPS, and SSL/TLS certificates should be kept up to date. API security extends beyond just authentication and authorization; it also involves protecting against rate limiting, brute-force attacks, and denial-of-service (DoS) attacks. Rate limiting restricts the number of requests a client can make within a specified timeframe, preventing abuse of the API. Brute-force attack prevention mechanisms should be in place to safeguard against unauthorized access attempts. Additionally, implementing rate limiting and access controls at the API gateway level can help mitigate DoS attacks. Secure handling of sensitive data is paramount in API security. Data encryption should be applied not only during transmission but also when storing sensitive information in databases or data stores. Encryption algorithms and key management should follow industry best practices. API endpoints should never return sensitive data, such as passwords or confidential user information, in response to client requests. Instead, APIs should return only the necessary information while protecting sensitive data from exposure. Monitoring and logging are essential components of API security. Regularly monitoring API traffic and analyzing logs can help detect unusual or suspicious activities. By setting up comprehensive logging, you can gain visibility into API usage patterns and potential security incidents. Logging should capture details of requests, responses, errors, and authentication events. Furthermore, implementing intrusion detection and prevention systems (IDPS) can help identify and respond to security threats in real-time. Security headers, such as Content Security Policy (CSP) and Cross-Origin Resource Sharing (CORS) headers, play a significant role in API security. CSP headers define which scripts and resources are allowed to execute on a web page, mitigating the risk of XSS attacks. CORS headers control which domains are permitted to make requests to the API, preventing unauthorized cross-origin requests. API versioning is another best practice that enhances security and compatibility. Versioning ensures that changes to the API do not break existing client applications. By using version numbers in API endpoints, you can introduce updates and new features while maintaining backward compatibility. Documentation is often overlooked but is a critical aspect of API security. Well-documented APIs provide developers with clear instructions on how to use the API, including authentication, authorization, and data formats. Comprehensive documentation can help developers use the API correctly and securely, reducing the likelihood of misconfigurations and vulnerabilities. API security is an ongoing process that requires regular updates and assessments. Keeping the API server and its dependencies up to date with security patches is essential. Additionally, conducting periodic security audits and code reviews can help identify and address security vulnerabilities in the API codebase. Lastly, educating both developers and users about API security *best practices is essential. Developers should receive training on secure coding practices, while users should be aware of the risks and security measures when interacting with APIs. In summary, API security is a multifaceted discipline that involves authentication, authorization, data validation, encryption, and more. By following these best practices and staying vigilant, you can enhance the security of your APIs and protect your applications and data from a wide range of threats.*

Chapter 9: Advanced Burp Suite Automation

Creating custom scripts in Burp Suite is an essential skill for security professionals looking to automate and extend the capabilities of this powerful tool. Custom scripts allow you to tailor Burp Suite's functionality to your specific testing needs and workflows, saving time and improving efficiency. Whether you're performing web application security assessments, penetration testing, or bug hunting, having the ability to automate repetitive tasks and perform complex operations can greatly enhance your productivity. Burp Suite, a popular web application security testing tool, provides a versatile scripting environment that supports both Python and Ruby programming languages. This flexibility enables you to write custom scripts that interact with web applications, manipulate requests and responses, and automate various testing tasks. Before diving into creating custom scripts, it's important to have a solid understanding of Burp Suite's core features and functionality. Familiarize yourself with Burp Suite's proxy, scanner, repeater, intruder, and other tools, as these will be the building blocks for your custom scripts. Having a good grasp of web application security concepts and common vulnerabilities is also crucial, as your scripts will often target specific security issues. To get started with scripting in Burp Suite, open the tool and navigate to the "Extender" tab. Here, you'll find the "BApp Store" where you can download and manage Burp extensions, including your custom scripts. The

"BApp Store" provides access to a variety of pre-built extensions and scripts created by the Burp Suite community, which can serve as valuable resources and examples for your own scripting projects. When creating custom scripts, you'll typically work in the "Extensions" tab within the "Extender" tab. Here, you can manage your scripts, load them into Burp Suite, and interact with the scripting console. Burp Suite's scripting environment allows you to write scripts for various purposes, such as automating repetitive tasks, customizing the scanning process, and exploiting security vulnerabilities. One common use case for custom scripts is to automate the identification of web application vulnerabilities. For example, you can write a script that scans a list of target URLs for common security issues like SQL injection or cross-site scripting (XSS). This script can send requests to each URL, analyze the responses for signs of vulnerability, and generate detailed reports. By automating vulnerability scanning, you can save time and ensure comprehensive coverage of your target applications. Another valuable application of custom scripts is in the exploitation of security vulnerabilities. Once you've identified a vulnerability, you can create a script that exploits it to demonstrate the impact and potential risks to the application owner. For instance, if you've found a SQL injection vulnerability, you can write a script that extracts sensitive data from the database or performs malicious actions. This not only helps in verifying the severity of the issue but also assists in crafting effective remediation recommendations. Custom Burp Suite scripts can also assist in performing *advanced attacks, such as session fixation, authentication bypass, and privilege escalation. By automating these attacks, you can thoroughly test the security of web applications and uncover potential weaknesses that might be missed with manual testing alone. Beyond vulnerability identification and exploitation, custom scripts can be used to automate tedious tasks, such as input fuzzing and parameter manipulation. For example, you can create a script that fuzzes input fields with various payloads to discover potential injection points and security flaws. Additionally, you can write scripts that modify request parameters on the fly, allowing you to test different input combinations and scenarios quickly. When developing custom scripts in Burp Suite, it's essential to maintain a structured and organized approach. Begin by defining the objectives and requirements of your script, outlining the tasks it needs to perform and the expected outcomes. Consider the inputs and outputs of your script, as well as any dependencies or external resources it may require. Next, choose the appropriate scripting language for your project, either Python or Ruby. Both languages are well-supported in Burp Suite and offer extensive libraries and modules for web application testing and automation. Once you've selected a scripting language, start writing your script incrementally, testing and debugging it as you go. Use the scripting console in Burp Suite to interactively develop and test small code snippets before integrating them into your script. As you develop your script, be mindful of error handling and exception management to ensure that it gracefully handles unexpected situations and errors. Furthermore, consider implementing logging and reporting mechanisms to track the script's progress and results. Burp Suite provides functions and libraries for logging messages and generating reports, making it easier to analyze and share the outcomes of your script executions. It's worth noting that Burp Suite offers extensive documentation and resources for scripting, including tutorials, guides, and a vibrant user community. Take advantage of these resources to learn more about scripting in Burp Suite and to seek assistance when encountering challenges or roadblocks. Moreover, consider contributing to the Burp Suite community by sharing your own custom scripts and extensions, helping others enhance their web application security testing capabilities. In summary, creating custom scripts in Burp Suite is a valuable skill that empowers security professionals to automate and extend the capabilities of this versatile tool. Custom scripts can be used for various purposes, including vulnerability identification, exploitation, advanced attacks, and task automation. By following best practices, maintaining a structured approach, and leveraging available resources, you can develop effective custom scripts that enhance your web application security assessments and testing workflows. Automating repetitive tasks with Burp Suite is a game-changer for security professionals, offering a way to streamline and accelerate various aspects of web application testing. By harnessing the power of automation, you can significantly reduce manual effort, improve efficiency, and enhance the overall effectiveness of your security assessments. Burp Suite, a leading web application security testing tool, provides robust automation capabilities through its extensible and scriptable architecture. The ability to automate tasks such as scanning, crawling, and data extraction not only saves time but also allows you to discover vulnerabilities more comprehensively. One of the fundamental tasks that can be automated in Burp Suite is the web application scanning process. Automated scanning involves sending a series of HTTP requests to a target web application and analyzing the responses for security vulnerabilities. Burp Suite's scanner tool automates this process by intelligently identifying potential vulnerabilities, such as SQL injection, Cross-Site Scripting (XSS), and more. Security professionals can configure the scanner to target specific areas of the application, define custom payloads, and set scan policies to tailor the scanning process to their needs. By automating scanning, you can quickly identify vulnerabilities and focus your manual testing efforts on more complex issues. Another critical aspect of web application testing that can be automated is the crawling of web pages. Crawling involves navigating through the application's structure to discover all accessible pages and endpoints. Burp Suite's crawler is designed to automate this process by following links and analyzing the application's responses to build a comprehensive map of the application. Automated crawling helps you understand the application's structure and ensures that no pages are overlooked during testing. Additionally, it aids in identifying hidden or less commonly used functionality that may contain security vulnerabilities. Automated data extraction is another powerful capability of Burp Suite. During security assessments, it's often necessary to extract data from web pages for further analysis or validation. Burp Suite's automated data extraction tools allow you to define extraction rules based on patterns, tags, or regular expressions. These rules enable you to capture specific information, such as email addresses, user names, or session tokens, from web pages automatically. The extracted data can then be used for testing or reporting purposes, saving you the time and effort required for manual data collection. Beyond scanning, crawling, and data extraction, Burp Suite's automation capabilities extend to other repetitive tasks, such as session management, parameter manipulation, and request/response modification. For example, you can automate the process of testing different input values for specific parameters by creating scripts that iterate through a list of payloads and automatically submit requests. Automation scripts can also be employed to manage sessions, perform authentication testing, and automate actions like logging in and out of web applications. By scripting these tasks, you can execute them consistently and repeatedly, ensuring thorough testing coverage. Automation in Burp Suite is achieved using its built-in support for scripting languages like Python and Ruby. Both languages offer extensive libraries and modules for web application testing and automation, making it possible to create custom automation scripts tailored to your unique testing needs. When developing automation scripts, it's essential to have a clear understanding of the web application's functionality, potential vulnerabilities, and the desired testing objectives. Start by defining the scope of your automation, including the specific tasks you intend to automate and the expected outcomes. Consider the inputs and outputs of your scripts, as well as any dependencies on external resources or data sources. It's important to script with care, ensuring that your automation actions align with ethical hacking and responsible testing practices. While automation can significantly improve efficiency, it should not compromise the integrity of your assessments or disrupt the target application's functionality. Throughout the development of automation scripts, thorough testing and validation are crucial. Burp Suite provides a scripting console that allows you to interactively test and debug your scripts, ensuring they work as intended. Pay attention to error handling and exception management in your scripts, as they play a vital role in gracefully handling unexpected situations and errors that may arise during automation. Moreover, leverage Burp Suite's logging and reporting capabilities to track the progress and results of your automated tasks. Logging helps you monitor script execution and identify any issues or anomalies. Furthermore, reporting allows you to document the outcomes of automation, making it easier to communicate findings and share results with stakeholders. It's worth noting that Burp Suite's extensible architecture provides access to a wide range of community-developed and third-party extensions that can enhance your automation capabilities. These extensions offer pre-built automation tools and scripts that can be integrated into your testing workflows, saving you time and effort. Additionally, Burp Suite offers comprehensive documentation, tutorials, and user communities where you can seek guidance and share insights on automation techniques. In summary, automating repetitive tasks with Burp Suite is a valuable skill for security professionals, enabling them to work more efficiently and effectively during web application testing. Automation can enhance scanning, crawling, data extraction, session management, and various other aspects of the assessment process. By scripting automation tasks with care and following best practices, security professionals can improve the accura*

Chapter 10: Reporting and Post-Exploitation

Generating comprehensive security reports is a critical step in any web application security assessment. These reports provide a clear and detailed overview of the security findings, vulnerabilities, and assessment results. A well-structured security report is not only essential for documenting the assessment but also for communicating the identified risks to stakeholders effectively. Burp Suite, a popular web application security testing tool, offers robust features for generating comprehensive security reports. These reports serve as valuable artifacts that demonstrate the security posture of the tested web application and provide actionable insights for remediation. When generating security reports with Burp Suite, it's essential to have a well-defined process in place to ensure accuracy and consistency. Start by accessing the reporting functionality within Burp Suite, typically found in the "Target" tab under the "Site map" section. Here, you can configure various options and settings to customize the content and format of the report. One of the first decisions to make when generating a security report is selecting the scope of the assessment. You can choose to generate a report for a specific target, a single scan, or an entire project encompassing multiple scans and assessments. Selecting the appropriate scope ensures that the report focuses on the relevant findings and assessment context. Next, consider the format of the report. Burp Suite offers several report templates, each tailored to different purposes and audiences. Common report formats include HTML, PDF, and XML. Choose the format that best aligns with your reporting requirements and audience preferences. Burp Suite also allows you to customize report templates, enabling you to include your organization's branding and logo for a professional look. Once you've defined the scope and format, you can configure the report's content. Burp Suite provides flexibility in selecting the findings and vulnerabilities to include in the report. You can filter the findings based on severity, status, or other criteria to focus on the most critical issues or provide a comprehensive overview. Additionally, you can choose to include evidence, such as HTTP request and response data, screenshots, and other artifacts, to support your findings and recommendations. Incorporating evidence enhances the report's credibility and helps stakeholders understand the context of identified vulnerabilities. It's important to include an executive summary at the beginning of the report. This summary provides a high-level overview of the assessment, including key findings, recommendations, and the overall risk assessment. The executive summary serves as a quick reference for stakeholders who may not have the time to review the entire report in detail. As you progress through the report configuration, consider including an introduction that provides context about the assessment's purpose, scope, and methodology. This introduction helps readers understand the assessment's objectives and the criteria used to evaluate the web application's security. In *addition to the introduction, it's beneficial to include a methodology section that outlines the testing approach, tools used, and any special considerations during the assessment. This methodology section offers transparency and allows stakeholders to assess the rigor of the testing process. When documenting vulnerabilities in the report, provide clear and concise descriptions of each security issue. Include information about the affected components, the attack vectors, potential impact, and any exploitation prerequisites. Use standardized naming conventions and severity ratings to ensure consistency and clarity in your findings. For each vulnerability, consider including remediation recommendations that guide the application owner on how to fix the issue. Recommendations should be actionable, specific, and prioritized based on risk to help the owner address the most critical vulnerabilities first. In addition to individual findings, it's valuable to include an overall risk assessment or risk matrix that categorizes the identified vulnerabilities by severity. This matrix helps stakeholders quickly grasp the overall security posture of the web application and make informed decisions about risk mitigation. When presenting findings, use clear and concise language to communicate technical details to non-technical stakeholders effectively. Avoid jargon and provide explanations that are accessible to a broader audience. In addition to text descriptions, consider using graphical elements such as charts and graphs to visualize data and trends. Charts can help highlight the distribution of vulnerabilities by severity, the prevalence of specific issue types, and the progress of remediation efforts over time. Furthermore, Burp Suite allows you to export scan data in various formats, such as CSV and JSON, for further analysis or integration with other security tools. This flexibility enables you to leverage scan data beyond the report and incorporate it into your organization's security processes. Throughout the report generation process, prioritize clarity and accuracy in your documentation. Proofread and review the report thoroughly to eliminate errors, inconsistencies, and ambiguities. A well-structured and error-free report enhances its credibility and readability. Once the report is generated, it's essential to distribute it to the relevant stakeholders promptly. Share the report with the application owner, development team, and other individuals responsible for addressing the identified vulnerabilities. Engage in discussions with stakeholders to ensure they understand the findings, the associated risks, and the recommended remediation actions. Effective communication is key to facilitating the resolution of security issues and promoting a collaborative approach to web application security. After distributing the report, follow up on the progress of remediation efforts and document any changes or updates to the findings. Maintain a record of the assessment and its outcomes for future reference and audit purposes. Burp Suite's reporting capabilities provide a valuable resource for web application security professionals, helping them document and communicate security assessments effectively. By following a structured report generation process, security practitioners can deliver comprehensive reports that facilitate informed decision-making and contribute to improved web application security. In summary, generating comprehensive security reports with Burp Suite is a crucial step in web application security assessments. These reports play a vital role in documenting findings, communicating risks, and guiding remediation efforts. By configuring and customizing reports to meet specific requirements and leveraging Burp Suite's reporting features effectively, security professionals can deliver reports that add value to their organizations and help secure web applications. Post-exploitation techniques and cleanup are critical aspects of ethical hacking and security testing that often receive less attention than the initial phases of an assessment. Once you've successfully identified and exploited vulnerabilities in a target system, it's essential to maintain access, gather valuable information, and ensure that your activities are discreet. This chapter explores the various post-exploitation techniques you can employ to maximize the value of your assessment while maintaining the highest ethical standards. Post-exploitation begins after an initial compromise of the target system, such as gaining unauthorized access or achieving code execution. At this stage, you typically have a foothold within the system and aim to expand your control while minimizing the risk of detection. One crucial post-exploitation objective is to establish persistence on the compromised system. Persistence involves ensuring that your access to the target system remains intact even after a reboot or system changes.*

Common persistence mechanisms include creating backdoors, modifying startup scripts, or leveraging rootkits to maintain control. The choice of persistence technique depends on the system's characteristics and the level of access you've achieved. Beyond persistence, post-exploitation activities also focus on information gathering. Once you have access, it's essential to extract valuable data, such as user credentials, configuration files, sensitive documents, and system logs. This information can be crucial for demonstrating the extent of the compromise and identifying potential further attack vectors. Various tools and techniques, including file extraction, memory analysis, and network reconnaissance, can aid in gathering this valuable data. Privilege escalation is another key post-exploitation objective. Even if you've initially compromised a low-privileged user, privilege escalation techniques can help you gain administrator or root-level access. Common privilege escalation methods include exploiting known vulnerabilities, misconfigurations, or weak access controls. To minimize the risk of detection, ethical hackers often employ "living off the land" techniques during post-exploitation. These techniques involve using built-in system utilities and legitimate processes to conduct malicious activities, making it more challenging for security monitoring tools to detect anomalous behavior. For example, PowerShell or Windows Management Instrumentation (WMI) can be leveraged for various post-exploitation tasks without raising suspicion. Maintaining stealth during post-exploitation is crucial to avoid triggering security alarms and alerts.

Security monitoring solutions and intrusion detection systems are often on high alert after an initial compromise, making it essential to obfuscate your activities. This can involve tampering with logs, evading security solutions, and minimizing network traffic associated with your activities. To further conceal your presence, it's advisable to remove or sanitize any forensic evidence that may lead investigators back to you or expose your tactics. Effective cleanup involves erasing traces of your actions, such as log entries, temporary files, and any custom scripts or tools you may have deployed. Ensure that you leave the system in a state that is as close as possible to its pre-compromise condition. Additionally, post-exploitation activities may include lateral movement within the network to explore and compromise other systems. Lateral movement techniques may involve exploiting vulnerabilities or weak credentials on neighboring systems to expand your access and control. However, it's crucial to exercise extreme caution during lateral movement to avoid disrupting network operations or causing unintended damage. Another critical consideration in post-exploitation is the ethical aspect of your actions. Maintaining the highest ethical standards is essential throughout the assessment, and this extends to post-exploitation activities. Ethical hackers should always operate within the boundaries defined by their scope and rules of engagement. Unnecessary damage, data theft, or other malicious activities are strictly prohibited. Additionally, it's essential to obtain explicit permission from the target organization for each post-exploitation action you plan to take. Even when you have initial access, you should not assume that all actions are fair game without proper authorization. The goal of post-exploitation is not to disrupt or damage systems but to assess their security posture, identify vulnerabilities, and assist in remediation efforts. During post-exploitation, ethical hackers often collaborate closely with the organization's security team to share findings, offer guidance, and ensure that vulnerabilities are addressed promptly. This collaborative approach helps organizations improve their security defenses and reduce the risk of future compromises. Moreover, it's crucial to maintain clear and detailed records of all post-exploitation activities. These records serve as evidence of your actions, your findings, and your recommendations for remediation. Effective documentation is vital not only for reporting but also for legal and compliance purposes. In some cases, organizations may need to report security breaches to regulatory authorities, and accurate records are essential for compliance with data protection laws. Ethical hackers should also consider the potential for countermeasures during post-exploitation. Target organizations may become aware of your activities and take steps to detect and respond to the intrusion. Being prepared for these scenarios involves planning exit strategies and ensuring that you can quickly and securely remove any persistent access or tools from the compromised systems. Post-exploitation cleanup is not only about erasing traces but also about ensuring that no unauthorized access remains. In summary, post-

exploitation techniques and cleanup are essential phases in ethical hacking and security testing. These activities involve maintaining access, gathering information, establishing persistence, and operating with discretion. Ethical hackers must operate within the defined scope, follow ethical guidelines, and collaborate with the target organization's security team to improve overall security.

BOOK 3
PENETRATION TESTING BEYOND WEB
NETWORK, MOBILE & CLOUD WITH BURP SUITE

ROB BOTWRIGHT

Chapter 1: Expanding Your Penetration Testing Horizons

The field of penetration testing, often referred to as ethical hacking, has evolved significantly over the years. In its early days, penetration testing was primarily a manual and ad-hoc process, where security professionals attempted to identify vulnerabilities in computer systems through trial and error. The focus was on finding and exploiting known vulnerabilities, and the tools available were limited in scope and functionality. However, as technology has advanced, so too has the practice of penetration testing. Today, penetration testing has become a structured and systematic approach to assessing the security of computer systems, networks, and applications. The landscape of penetration testing has expanded to encompass a wide range of techniques, tools, and methodologies. One of the significant changes in the field is the shift from manual testing to a more automated approach. Automation has allowed security professionals to conduct more comprehensive and efficient assessments, reducing the time and effort required for testing. Modern penetration testing tools, such as Burp Suite, Metasploit, and Nessus, offer powerful capabilities for vulnerability scanning, exploitation, and reporting. These tools have become indispensable for security professionals, enabling them to identify and address vulnerabilities quickly and accurately. Another notable development is the increasing focus on web application security testing. As organizations increasingly rely on web applications for their business operations, web application vulnerabilities have become a primary target for attackers. Penetration testers have adapted to this trend by specializing in web application testing and using tools specifically designed for this purpose. The demand for web application security expertise has grown, and organizations are investing more in securing their web applications. Furthermore, penetration testing has expanded beyond the traditional network perimeter. With the rise of cloud computing and mobile technologies, the attack surface has expanded, and penetration testers must adapt to assess security in these new environments. Testing cloud-based infrastructure, mobile applications, and Internet of Things (IoT) devices has become essential to providing comprehensive security assessments. Additionally, the regulatory landscape has evolved, leading to increased compliance requirements for organizations. Penetration testers must now consider compliance frameworks such as the Payment Card Industry Data Security Standard (PCI DSS) and the General Data Protection Regulation (GDPR) when conducting assessments. These regulations have introduced new challenges and considerations for penetration testing professionals. One of the most significant shifts in penetration testing is the emphasis on collaboration and communication. Penetration testers are no longer seen as isolated individuals conducting assessments in isolation; instead, they work closely with organizations to understand their specific needs and goals. This collaboration allows penetration testers to tailor their assessments to the organization's unique environment and priorities. Moreover, organizations have recognized the importance of integrating penetration testing into their broader cybersecurity programs. Regular assessments and continuous monitoring have become essential components of a proactive cybersecurity strategy. As a result, penetration testing has moved from a one-time event to an ongoing process. Penetration testers now engage with organizations in a continuous feedback loop, helping them identify and address vulnerabilities as they emerge. This shift towards continuous testing aligns with the broader concept of DevSecOps, where security is integrated into the software development lifecycle from the beginning. Another notable change is the increasing focus on social engineering and human factors in penetration testing. Attackers often target individuals within organizations through tactics such as phishing and pretexting. Penetration testers have recognized the need to assess the human element of security and develop strategies to educate and train employees to recognize and respond to social engineering attacks. Additionally, the emergence of bug bounty programs and crowdsourced security testing platforms has created new opportunities for penetration testers. Organizations now incentivize ethical hackers to find and report vulnerabilities in their systems, offering monetary rewards and recognition for their efforts. These programs have expanded the pool of talent available for security testing and created a more diverse and dynamic testing ecosystem. The evolving landscape *of penetration testing has also led to changes in the skill sets and qualifications required of security professionals. Certifications such as the Certified Ethical Hacker (CEH) and the Offensive Security Certified Professional (OSCP) have become highly regarded in the industry. These certifications validate the knowledge and skills of penetration testers and provide a standardized way to assess their capabilities. Furthermore, penetration testers must stay updated on the latest threats, vulnerabilities, and attack techniques. Continuous learning and professional development are essential to remain effective in the field. In summary, the landscape of penetration testing has evolved significantly, driven by advances in technology, changes in regulatory requirements, and the growing awareness of cybersecurity threats. Today, penetration testing is a dynamic and collaborative process that encompasses a wide range of techniques, tools, and methodologies. Penetration testers play a crucial role in helping organizations identify and address vulnerabilities, ultimately strengthening their cybersecurity defenses. In the world of penetration testing, the choice of targets is a critical decision that significantly influences the testing process and the potential impact on an organization's security posture. Targets can vary widely, and each type presents unique challenges and considerations for penetration testers. Let's explore some of the different types of targets that ethical hackers may encounter during their assessments. One common type of target is the web application, which has become a primary focus for penetration testers in recent years. Web applications are ubiquitous, and they often serve as a gateway to sensitive data and systems. Assessing the security of web applications involves identifying vulnerabilities such as SQL injection, cross-site scripting (XSS), and insecure authentication mechanisms. Web application penetration testing requires a deep understanding of web technologies, protocols, and common coding vulnerabilities. Another type of target is the network infrastructure, which includes routers, switches, firewalls, and other devices that control the flow of data within an organization's network. Assessing the security of network infrastructure involves identifying weaknesses in configuration, access control, and potential points of entry for attackers. Penetration testers may attempt to exploit vulnerabilities in network devices to gain unauthorized access or disrupt network operations. Mobile applications are another type of target that has gained prominence with the widespread use of smartphones and tablets. Mobile app penetration testing focuses on identifying vulnerabilities specific to mobile platforms, such as insecure data storage, insecure communication, and mobile app-specific authentication issues. Testers must also consider the security of the backend APIs that mobile apps interact with. Penetration testers often assess cloud-based environments, including Infrastructure as a Service (IaaS), Platform as a Service (PaaS), and Software as a Service (SaaS) offerings. Cloud security assessments involve evaluating the configuration of cloud resources, identifying misconfigurations, and assessing the security of data stored in the cloud. Cloud providers offer various security features and controls that testers must navigate to uncover vulnerabilities. IoT devices, such as smart thermostats, security cameras, and industrial sensors, represent a unique and growing target for penetration testers. These devices often lack robust security features and may expose vulnerabilities that could be exploited to gain control over them or compromise the network they are connected to. IoT penetration testing requires knowledge of both hardware and software vulnerabilities. Network penetration testing is a broad category that encompasses various types of networks, including corporate networks, data centers, and industrial control systems (ICS). Penetration testers assess the security of these networks by identifying vulnerabilities, exploiting weaknesses, and attempting to gain unauthorized access to critical systems. ICS environments, in particular, present unique challenges due to their specialized nature. Social engineering assessments focus on the human element of security. Penetration testers may attempt to manipulate individuals within an organization to disclose sensitive information, click on malicious links, or perform actions that could compromise security. Social engineering assessments often involve tactics such as phishing, pretexting, and tailgating. External perimeter testing involves assessing an organization's external-facing systems, such as web servers, email servers, and DNS servers. Testers attempt to identify vulnerabilities that could be exploited from the internet to gain access or disrupt services. This type of testing may also include* evaluating the effectiveness of intrusion detection and prevention systems. Internal network testing focuses on assessing the security of an organization's internal network, often assuming that an attacker has gained a foothold within the network. Testers aim to identify lateral movement opportunities and weaknesses in internal security controls. Wireless network assessments involve evaluating the security of wireless networks, including Wi-Fi networks and Bluetooth devices. Testers look for vulnerabilities such as weak encryption, misconfigured access points, and rogue devices. Wireless assessments may also include testing the security of mobile device connections. Physical security assessments assess the physical security measures in place to protect an organization's assets. Testers may attempt to gain physical access to facilities, bypass access control systems, or tamper with physical security mechanisms. These assessments help organizations identify weaknesses in their physical security controls. Cloud security assessments involve evaluating the configuration of cloud resources, identifying misconfigurations, and assessing the security of data stored in the cloud. Cloud providers offer various security features and controls that testers must navigate to uncover vulnerabilities. IoT devices, such as smart thermostats, security cameras, and industrial sensors, represent a unique and growing target for penetration testers. These devices often lack robust security features and may expose vulnerabilities that could be exploited to gain control over them or compromise the network they are connected to. IoT penetration testing requires *knowledge of both hardware and software vulnerabilities. In summary, penetration testers must be prepared to assess a wide range of targets, each with its unique characteristics and challenges. The choice of target depends on the organization's goals, the systems and technologies in use, and the specific risks that need to be addressed. Ethical hackers use their expertise and tools to uncover vulnerabilities, provide recommendations for remediation, and ultimately contribute to improving an organization's security posture.*

Chapter 2: Setting Up Burp Suite for Diverse Targets

Customizing Burp Suite for network assessments is an essential skill that allows penetration testers to tailor the tool to the specific needs of the engagement. Burp Suite, a powerful web application security testing tool, offers extensive customization options that can be leveraged to assess network infrastructure, network devices, and services effectively. Network assessments often involve evaluating the security of routers, switches, firewalls, and other critical network components. Customizing Burp Suite for network assessments starts with understanding the tool's capabilities and features. Burp Suite is known for its versatility in web application testing, but it can also be employed for network penetration testing. The tool's proxy functionality, in particular, can be adapted for network assessments by configuring it to intercept and analyze network traffic. To get started, penetration testers need to configure Burp Suite's proxy listener to listen on the appropriate network interface and port. This allows Burp Suite to capture and analyze traffic passing through the designated network segment. Once the proxy listener is set up, testers can configure their target devices or network services to route their traffic through the Burp Suite proxy. This redirection of traffic enables testers to intercept and inspect network packets, much like they would with web requests and responses in web application testing. Burp Suite *provides a user-friendly interface for viewing and manipulating intercepted network traffic. Testers can examine the contents of packets, including headers and payloads, to identify vulnerabilities, misconfigurations, or unusual behavior. Additionally, Burp Suite's extensibility through extensions and scripts opens up opportunities for advanced customization. Penetration testers can develop or leverage existing extensions that provide specialized functionality for network assessments. For example, extensions can be used to automate common tasks, perform specific tests, or integrate with other tools and services. Custom extensions can be written in various programming languages, such as Python or Java, and integrated seamlessly into Burp Suite. Network assessments often involve assessing the security of network protocols and services. Custom Burp Suite extensions can be designed to interact with these protocols, allowing testers to conduct comprehensive assessments. Another aspect of customizing Burp Suite for network assessments is configuring its built-in tools and features to suit the engagement's requirements. For example, testers can leverage Burp Suite's Intruder tool to perform brute-force attacks, fuzzing, or other automated tests on network services. By tailoring the payload lists and attack parameters to the target service, testers can identify weak credentials, vulnerabilities, or unexpected behavior. Additionally, Burp Suite's Repeater tool can be used to interact with network services interactively. Testers can craft and send specific requests to services, observing how they respond and identifying potential* security issues. Furthermore, Burp Suite's scanner functionality can be utilized to automate the discovery of vulnerabilities in network devices and services. By configuring scan policies, testers can instruct Burp Suite to perform targeted scans based on the specific network assessment goals. This can include identifying known vulnerabilities in network device firmware, checking for weak configurations, or detecting potential security misconfigurations. To enhance the effectiveness of network assessments, penetration testers can leverage Burp Suite's reporting capabilities. Customized reports can be generated to document findings, including vulnerabilities, their impact, and recommended remediation steps. Reports can be tailored to the audience, whether it's technical teams responsible for implementing fixes or executive stakeholders seeking an overview of security risks. Burp Suite supports various report formats, allowing testers to choose the most suitable format for their client or organization. As with any penetration testing engagement, collaboration and communication play a crucial role in network assessments. Customized reports should provide clear and concise information that facilitates discussions between the testing team and the client's IT and security teams. Effective communication ensures that identified issues are understood, prioritized, and addressed promptly. In summary, customizing Burp Suite for network assessments empowers penetration testers to adapt this versatile tool to the specific requirements of evaluating network infrastructure, devices, and services. By configuring proxy listeners, developing custom *extensions, and fine-tuning tool settings, testers can conduct comprehensive and effective network assessments. Burp Suite's flexibility and extensibility make it a valuable asset for identifying vulnerabilities, misconfigurations, and security risks in network environments. Through well-documented reports and clear communication, penetration testers can contribute to improving an organization's network security posture and mitigating potential threats. Adapting Burp Suite for mobile application testing is a crucial skill for penetration testers, as mobile apps have become an integral part of our daily lives. Mobile devices store vast amounts of sensitive information, and ensuring the security of mobile applications is paramount. Burp Suite, known for its prowess in web application security testing, can be tailored for mobile app assessments with some customization and understanding of mobile-specific challenges. To begin adapting Burp Suite for mobile testing, testers need to set up a proxy configuration on both their mobile device and the Burp Suite tool. This configuration allows Burp Suite to intercept and analyze the HTTP(S) traffic between the mobile app and its backend servers. Once the proxy is configured, testers can route their mobile device's traffic through Burp Suite by specifying the tool's IP address and port as the proxy settings on the device. This interception enables testers to observe and manipulate the network traffic generated by the mobile app, similar to how they would inspect web requests in traditional web application testing. It's important to note that mobile app testing involves assessing both the* client-side (the app itself) and the server-side (the backend services) components. Burp Suite can be used to examine and tamper with the requests and responses between the mobile app and the server, revealing potential vulnerabilities and security weaknesses. One key aspect of mobile application security testing is the examination of how the app stores and handles sensitive data. Mobile apps often deal with user credentials, personal information, and financial data, making it crucial to identify and assess how this data is stored and transmitted. Burp Suite's proxy and interception capabilities come into play as testers analyze the data flows and look for security gaps. While the interception of network traffic is essential for mobile app testing, testers should also consider other mobile-specific attack vectors. One such vector is the analysis of the mobile app's binary code. Mobile app binaries, which contain the compiled code and resources of the app, can be reverse-engineered to discover vulnerabilities or hardcoded secrets. Testers can use tools like JADX, APKTool, and Frida to decompile, disassemble, and analyze the app's code and behavior. By gaining insights into the app's logic, testers can identify potential weaknesses or insecure coding practices. Additionally, mobile app assessments often include testing the app's interaction with the device's hardware features, such as the camera, microphone, GPS, and sensors. Burp Suite can assist in assessing how the app requests and utilizes these permissions. For example, testers can analyze requests related to location data or multimedia access to ensure that the app requests permissions *appropriately and does not misuse them. Another critical aspect of mobile app testing is identifying and testing the security of backend APIs that the app communicates with. API endpoints may expose sensitive functionality or data that attackers could target. Testers can use Burp Suite to analyze API requests and responses, searching for vulnerabilities like insecure authentication, insufficient authorization checks, or API endpoint vulnerabilities. Burp Suite's Intruder tool can be customized to automate API testing by sending a series of requests with varying inputs and analyzing the responses for anomalies. Mobile app assessments may also involve testing for potential vulnerabilities introduced by third-party libraries or SDKs used by the app. Testers can use Burp Suite to inspect the network traffic generated by these libraries and identify any security concerns. In some cases, mobile app testing may require testers to bypass SSL/TLS pinning, a security feature used by apps to prevent interception of SSL/TLS-encrypted traffic. Burp Suite allows testers to decrypt and inspect the traffic even when SSL/TLS pinning is implemented, aiding in identifying vulnerabilities. Additionally, Burp Suite offers extensions and customizations that can be employed to automate repetitive tasks and enhance testing efficiency. Custom scripts can be written to extend Burp Suite's capabilities for mobile app testing, addressing specific challenges or requirements. In summary, adapting Burp Suite for mobile application testing empowers penetration testers to evaluate the security of mobile apps comprehensively. By intercepting network traffic, analyzing binary code, assessing hardware interactions, examining API security, and considering third-party components, testers can uncover vulnerabilities and provide recommendations for remediation. Effective mobile app testing requires a blend of tools, techniques, and a deep understanding of mobile-specific security issues, all of which Burp Suite can facilitate when appropriately configured and customized. As mobile technology continues to advance, ensuring the security of mobile applications remains an ongoing challenge, making thorough testing and customization with tools like Burp Suite more critical than ever.*

Chapter 3: Network Penetration Testing with Burp Suite

Scanning and profiling network services are fundamental steps in the process of assessing the security of a network, whether for penetration testing, vulnerability assessment, or general network hygiene. These activities aim to identify open ports, running services, and potential vulnerabilities that may exist on network devices. Network scanning involves the use of specialized tools and techniques to gather information about a target network's configuration and the services running on it. One of the most commonly used tools for network scanning is Nmap (Network Mapper), a versatile and powerful open-source tool that provides a wide range of scanning and profiling capabilities. Nmap allows testers to discover devices, map network topology, and identify services, operating systems, and potential vulnerabilities. To initiate a network scan with Nmap, testers typically specify the target IP address or range and the desired scan options. Nmap supports various scan types, such as the SYN scan, which sends TCP SYN packets to probe for open ports, and the UDP scan, which focuses on UDP-based services. Each scan type has its advantages and limitations, and the choice depends on the specific testing objectives and network characteristics. Network profiling, on the other hand, goes beyond identifying open ports and services; it aims to gather detailed information about those services and their configurations. Profiling is particularly useful for understanding how services are implemented and assessing their potential vulnerabilities. One of the techniques used in service profiling is banner grabbing, which involves connecting to open ports and extracting information from the service banners that are often provided. These banners may reveal the service version, application name, and sometimes even additional details about the system. Banner grabbing can be performed using tools like Telnet, Netcat, or specialized banner-grabbing scripts. By understanding the specific service versions and configurations, testers can look for known vulnerabilities associated with those versions and tailor their tests accordingly. Another aspect of network service profiling is actively querying services for information. For example, testers can use the Simple Network Management Protocol (SNMP) to query network devices for system information, interface statistics, and more. SNMP queries provide valuable insights into device configurations and performance, allowing testers to assess the security of SNMP settings. Beyond SNMP, testers can use specialized tools like SMBclient to interact with Windows file shares, SMTP commands to query mail servers, and DNS queries to explore DNS configurations. Profiling also extends to web services, where testers can analyze HTTP responses for web applications. Web service profiling may involve examining HTTP headers, cookies, and web server banners to understand the technologies in use. Web vulnerability scanners, such as Burp Suite or OWASP ZAP, can be used to automate the discovery of potential web application vulnerabilities, further enhancing the *profiling process. Additionally, profiling may include checking for misconfigurations or weak security settings on network services. For example, testers can assess the strength of SSL/TLS configurations on HTTPS services, ensuring that secure protocols and encryption algorithms are in use. Identifying misconfigured services that may expose sensitive information or weaken security controls is crucial for overall network security. In some cases, network services may be accessible via default or weak credentials. Testers can conduct credential testing to determine whether default or common credentials can be used to gain unauthorized access to services. This involves using tools like Hydra or Medusa to perform brute-force or dictionary attacks against services like SSH, Telnet, or FTP. It's important to note that conducting network scans and service profiling must be done with proper authorization and consent. Unauthorized scanning can disrupt network operations and may be illegal in some jurisdictions. Testers should always obtain permission from the network owner or administrator before conducting any scanning or profiling activities. In summary, scanning and profiling network services are critical steps in evaluating the security of a network. These activities provide valuable insights into the network's configuration, open ports, running services, and potential vulnerabilities. Using tools like Nmap for scanning and banner grabbing, as well as actively querying services and checking for misconfigurations, testers can comprehensively assess the security posture of a network. The information gathered through scanning and profiling forms the foundation for further penetration testing, vulnerability assessment, and security improvement efforts. Exploring the topic of exploiting network vulnerabilities with Burp Suite is a fascinating journey into the realm of ethical hacking and penetration testing. Burp Suite, a versatile and powerful tool known for web application security testing, can also be leveraged to assess and exploit network vulnerabilities effectively. Before diving into the specifics of exploiting network vulnerabilities, it's crucial to understand what network vulnerabilities are and why they pose a significant risk. Network vulnerabilities are weaknesses or flaws in a network's configuration, protocols, or devices that can be exploited by malicious actors to gain unauthorized access, disrupt services, or steal sensitive information. These vulnerabilities can result from misconfigurations, outdated software, weak encryption, or design flaws in network devices and services. Exploiting these vulnerabilities allows ethical hackers to demonstrate the potential impact of security weaknesses and help organizations strengthen their defenses. One of the primary goals of exploiting network vulnerabilities is to gain unauthorized access to a target system or device. This access can be leveraged to further assess the system's security, gather sensitive information, or demonstrate the potential impact of a successful attack. Burp Suite's Intruder tool is a valuable asset in this context, as it allows testers to automate and customize attacks against network services. For example, testers can use Intruder to perform brute-force attacks on services like SSH or FTP, attempting to guess weak or* default credentials. Additionally, Intruder can be configured to conduct dictionary attacks, where a list of potential passwords is systematically tested to gain access. Another common objective of network vulnerability exploitation is to disrupt services or cause a denial of service (DoS) condition. DoS attacks aim to overwhelm a network or system, making it inaccessible to legitimate users. Burp Suite can assist in assessing a network's resilience against DoS attacks by generating and sending a high volume of traffic to target services. This traffic simulation allows testers to identify potential weaknesses in a network's ability to handle excessive requests and helps organizations fortify their defenses. Exploiting network vulnerabilities can also involve leveraging known vulnerabilities in network devices or software. Ethical hackers often use tools like Metasploit, which integrates seamlessly with Burp Suite, to exploit known vulnerabilities in a controlled environment. By demonstrating the exploitation of a known vulnerability, testers can highlight the importance of timely patching and security updates. Furthermore, ethical hackers can simulate the actions of malicious actors by exploiting network vulnerabilities to gain unauthorized access and escalate privileges. This process involves taking advantage of vulnerabilities to elevate the attacker's privileges within the network, ultimately allowing access to more critical systems and data. Burp Suite's extensive capabilities, including scanning, reconnaissance, and exploitation, make it a valuable tool for this purpose. Additionally, ethical hackers may utilize network vulnerability exploitation to demonstrate the potential *consequences of a successful attack, such as data exfiltration. Burp Suite's ability to intercept and manipulate network traffic allows testers to simulate data theft scenarios, showcasing the risks associated with security weaknesses. It's important to note that ethical hacking and network vulnerability exploitation should always be conducted with proper authorization and consent. Unauthorized exploitation of vulnerabilities can have legal and ethical implications, so testers must obtain permission from network owners or administrators before conducting any tests. Furthermore, network vulnerability exploitation should be approached with caution to avoid causing harm or damage to systems or networks. Ethical hackers must adhere to responsible disclosure practices, reporting their findings to organizations so that vulnerabilities can be remediated promptly. In summary, exploiting network vulnerabilities with Burp Suite is a valuable and necessary component of ethical hacking and penetration testing. It allows testers to assess the security posture of networks, demonstrate the potential impact of vulnerabilities, and help organizations strengthen their defenses. By using Burp Suite's Intruder tool, Metasploit integration, and traffic manipulation capabilities, ethical hackers can effectively assess and exploit network vulnerabilities while adhering to ethical and legal guidelines. The knowledge and insights gained through these activities contribute to improved network security and resilience against real-world threats.*

Chapter 4: Mobile Application Assessment with Burp Suite

Creating a mobile testing environment is a crucial step for conducting comprehensive security assessments of mobile applications. In today's digital landscape, mobile devices play a significant role in our daily lives, and mobile applications store and process a wealth of sensitive information. For this reason, it's essential for organizations to ensure the security of their mobile apps. Setting up a mobile testing environment involves creating a controlled and isolated environment where testers can assess the security of mobile applications without affecting production systems. This controlled environment allows testers to identify vulnerabilities, test for potential attacks, and provide recommendations for improving the security of the mobile application. One of the fundamental components of a mobile testing environment is the mobile device itself. Testers typically use physical smartphones and tablets or emulators to assess the security of mobile applications. Physical devices provide a more accurate representation of real-world scenarios, while emulators offer convenience and flexibility for testing various device configurations and versions. Testers should consider using a combination of physical devices and emulators to cover a wide range of testing scenarios. Additionally, it's important to choose mobile devices that are representative of the target audience, as different devices and operating system versions may exhibit unique security challenges. Once the mobile devices are selected, testers need to establish a dedicated network environment for mobile app testing. This network environment should be isolated from the production network to prevent any accidental impact on operational systems. Using a separate network or subnet, testers can control network traffic and simulate various network conditions to assess how the mobile app behaves in different scenarios. To monitor and analyze network traffic during mobile app testing, testers can employ network sniffing tools like Wireshark or specialized mobile app proxy tools like Burp Suite Mobile Assistant. These tools enable testers to intercept and inspect HTTP and HTTPS requests made by the mobile app, helping them identify potential security vulnerabilities. In addition to network isolation, testers should also consider implementing security controls like firewalls and intrusion detection systems (IDS) within the mobile testing environment to enhance security and detect potential threats. A critical aspect of mobile app security testing is the use of mobile application security testing frameworks. Frameworks like OWASP Mobile Security Testing Guide and the Mobile Security Framework (MobSF) provide a structured approach to assessing the security of mobile applications. These frameworks offer guidelines, checklists, and automated testing scripts to help testers systematically evaluate mobile app security. Testers should familiarize themselves with these frameworks and leverage their resources to ensure thorough testing. To install and configure these tools, testers can refer to their documentation and follow the provided instructions. In addition to using security testing frameworks, testers should also consider employing static and dynamic analysis tools. Static analysis tools examine the source code or binary of the mobile app without executing it, identifying potential security issues like code vulnerabilities and insecure coding practices. Dynamic analysis tools, on the other hand, run the mobile app in a controlled environment and monitor its behavior, identifying runtime security vulnerabilities. Effective use of static and dynamic analysis tools enhances the depth and coverage of mobile app security testing. To set up these tools, testers should follow the installation and configuration guidelines provided by the tool developers. Another important aspect of a mobile testing environment is the ability to simulate real-world scenarios. Testers should create test cases and scenarios that mimic user interactions with the mobile app. This includes testing for various inputs, actions, and conditions that users may encounter. By replicating real-world usage patterns, testers can identify vulnerabilities that could be exploited by malicious users. Furthermore, testers should consider using a variety of testing techniques, such as penetration testing, to assess the security of the mobile app. Penetration testing involves actively attempting to exploit vulnerabilities in the mobile app to determine their impact and potential consequences. This approach provides valuable insights into the app's resilience against real-world attacks. To carry out penetration testing, testers should have a thorough understanding of common mobile app *vulnerabilities and attack vectors. They should also follow ethical hacking practices and obtain proper authorization from the app owner or administrator. Additionally, testers should document their findings, including vulnerabilities and recommended remediation steps, to provide comprehensive feedback to the app development team. In summary, setting up a mobile testing environment is a critical prerequisite for conducting effective security assessments of mobile applications. This controlled environment, consisting of mobile devices, network isolation, security tools, testing frameworks, and realistic test scenarios, enables testers to identify and address security vulnerabilities in mobile apps. By following best practices and ethical hacking guidelines, testers can contribute to the improvement of mobile app security, protecting sensitive data and enhancing the overall security posture of organizations in an increasingly mobile-driven world. Analyzing and manipulating mobile app traffic is a fundamental aspect of mobile application security testing. In today's mobile-centric world, ensuring the security of mobile applications is crucial, as these apps handle sensitive user data and perform critical functions. Analyzing mobile app traffic allows testers to understand how the app communicates with remote servers and identify potential security vulnerabilities. One of the key tools for analyzing mobile app traffic is a mobile app proxy, such as Burp Suite Mobile Assistant. These proxies act as intermediaries between the mobile app and the server, intercepting and logging all traffic that passes through. By intercepting this traffic, testers gain visibility into the data exchanged between the mobile app and the server, including requests and responses. This visibility is invaluable for identifying security issues, understanding the app's behavior, and assessing its overall security posture. Mobile app proxies like Burp Suite Mobile Assistant offer features for intercepting and modifying requests and responses. Testers can set up intercept rules to selectively capture and inspect specific requests, allowing them to focus on critical parts of the app's communication. For instance, testers can intercept login requests to examine authentication mechanisms or intercept data upload requests to assess how user data is handled. Once intercepted, requests can be modified to test how the app handles unexpected input or to simulate different scenarios. For example, testers can change the content of a request to see if the app properly validates and sanitizes input data. They can also manipulate headers to simulate different network conditions or user-agent strings to test how the app responds to various devices. Beyond interception and modification, mobile app proxies provide powerful features for analyzing traffic, such as traffic logs and request/response views. These tools allow testers to review the entire communication flow between the app and the server, helping them identify anomalies, security weaknesses, or suspicious behavior. For instance, testers can use traffic logs to trace the sequence of requests and responses during app interactions, helping them understand how data flows within the app. Request/response views provide detailed information about each interaction, including headers, parameters,* and payloads. Testers can inspect these details to pinpoint security issues like insecure communication protocols or data leakage. In addition to analyzing mobile app traffic, testers can use mobile app proxies for more advanced testing techniques, such as security scanning. Mobile app proxies can integrate with security testing tools like OWASP ZAP to automate security scans. These scans identify common vulnerabilities like insecure transport, missing security headers, and improper input validation. By automating security scans, testers can quickly identify low-hanging fruit and prioritize their efforts on more complex security issues. Furthermore, mobile app proxies can be used to assess the app's resistance to various network attacks, such as man-in-the-middle (MITM) attacks. MITM attacks occur when an attacker intercepts and modifies communication between the mobile app and the server. Mobile app proxies enable testers to simulate MITM attacks and evaluate how the app responds to such threats. This assessment is crucial for ensuring that the app employs secure communication practices and can withstand real-world attacks. To perform MITM simulations, testers configure the mobile device to trust the proxy's root certificate, allowing the proxy to decrypt and inspect encrypted traffic. This setup is essential for analyzing HTTPS traffic, which is encrypted by default to protect sensitive data. By decrypting and inspecting HTTPS traffic, testers can identify potential security weaknesses and assess the app's encryption practices. While analyzing and manipulating mobile app traffic is a powerful method for uncovering security *vulnerabilities, testers must approach this activity responsibly and ethically. Testing mobile app traffic should be conducted with proper authorization and consent from the app owner or administrator. Unauthorized interception or manipulation of data is illegal and unethical. Testers must also handle any sensitive user data encountered during testing with the utmost care and confidentiality, ensuring that it is not exposed or mishandled. In summary, analyzing and manipulating mobile app traffic is a vital component of mobile application security testing. Mobile app proxies like Burp Suite Mobile Assistant provide the necessary tools and capabilities to intercept, inspect, and modify mobile app traffic, enabling testers to identify security vulnerabilities, assess communication practices, and evaluate the app's resilience against network attacks. By conducting this testing responsibly and ethically, testers play a critical role in enhancing the security of mobile applications, protecting user data, and mitigating potential risks in the mobile app ecosystem.*

Chapter 5: Securing Cloud Environments with Burp Suite

Analyzing and manipulating mobile app traffic is a fundamental aspect of mobile application security testing. In today's mobile-centric world, ensuring the security of mobile applications is crucial, as these apps handle sensitive user data and perform critical functions. Analyzing mobile app traffic allows testers to understand how the app communicates with remote servers and identify potential security vulnerabilities. One of the key tools for analyzing mobile app traffic is a mobile app proxy, such as Burp Suite Mobile Assistant. These proxies act as intermediaries between the mobile app and the server, intercepting and logging all traffic that passes through. By intercepting this traffic, testers gain visibility into the data exchanged between the mobile app and the server, including requests and responses. This visibility is invaluable for identifying security issues, understanding the app's behavior, and assessing its overall security posture. Mobile app proxies like Burp Suite Mobile Assistant offer features for intercepting and modifying requests and responses. Testers can set up intercept rules to selectively capture and inspect specific requests, allowing them to focus on critical parts of the app's communication. For instance, testers can intercept login requests to examine authentication mechanisms or intercept data upload requests to assess how user data is handled. Once *intercepted, requests can be modified to test how the app handles unexpected input or to simulate different scenarios. For example, testers can change the content of a request to see if the app properly validates and sanitizes input data. They can also manipulate headers to simulate different network conditions or user-agent strings to test how the app responds to various devices. Beyond interception and modification, mobile app proxies provide powerful features for analyzing traffic, such as traffic logs and request/response views. These tools allow testers to review the entire communication flow between the app and the server, helping them identify anomalies, security weaknesses, or suspicious behavior. For instance, testers can use traffic logs to trace the sequence of requests and responses during app interactions, helping them understand how data flows within the app. Request/response views provide detailed information about each interaction, including headers, parameters, and payloads. Testers can inspect these details to pinpoint security issues like insecure communication protocols or data leakage. In addition to analyzing mobile app traffic, testers can use mobile app proxies for more advanced testing techniques, such as security scanning. Mobile app proxies can integrate with security testing tools like OWASP ZAP to automate security scans. These scans identify common vulnerabilities like insecure transport, missing security headers, and improper input validation. By automating security scans, testers can quickly identify low-hanging fruit and prioritize their efforts on more complex security issues. Furthermore, mobile app proxies can be used to assess the app's resistance to various network attacks, such as man-in-the-middle (MITM) attacks. MITM attacks occur when an attacker intercepts and modifies communication between the mobile app and the server. Mobile app proxies enable testers to simulate MITM attacks and evaluate how the app responds to such threats. This assessment is crucial for ensuring that the app employs secure communication practices and can withstand real-world attacks. To perform MITM simulations, testers configure the mobile device to trust the proxy's root certificate, allowing the proxy to decrypt and inspect encrypted traffic. This setup is essential for analyzing HTTPS traffic, which is encrypted by default to protect sensitive data. By decrypting and inspecting HTTPS traffic, testers can identify potential security weaknesses and assess the app's encryption practices. While analyzing and manipulating mobile app traffic is a powerful method for uncovering security vulnerabilities, testers must approach this activity responsibly and ethically. Testing mobile app traffic should be conducted with proper authorization and consent from the app owner or administrator. Unauthorized interception or manipulation of data is illegal and unethical. Testers must also handle any sensitive user data encountered during testing with the utmost care and confidentiality, ensuring that it is not exposed or mishandled. In summary, analyzing and manipulating mobile app traffic is a vital component of mobile application security testing. Mobile app proxies like Burp Suite Mobile Assistant provide the necessary tools and capabilities to intercept, inspect, and modify* mobile app traffic, enabling testers to identify security vulnerabilities, assess communication practices, and evaluate the app's resilience against network attacks. By conducting this testing responsibly and ethically, testers play a critical role in enhancing the security of mobile applications, protecting user data, and mitigating potential risks in the mobile app ecosystem. Leveraging Burp Suite for cloud security audits is a strategic approach to assess and enhance the security of cloud environments. In an increasingly cloud-centric world, organizations must ensure the security of their data, applications, and infrastructure hosted in the cloud. Cloud security audits are essential for identifying vulnerabilities, misconfigurations, and compliance issues that may put your organization at risk. Burp Suite, a powerful web vulnerability scanner and proxy tool, can be a valuable asset in your arsenal for conducting thorough cloud security audits. While Burp Suite is primarily known for web application security testing, its capabilities extend beyond traditional web applications and can be adapted for cloud-specific assessments. To leverage Burp Suite effectively for cloud security audits, you need to understand its features and how to tailor them to your cloud environment. First and foremost, you should have a clear understanding of your cloud infrastructure's architecture and components. Identify the cloud services, APIs, and data storage solutions you use, as well as their interdependencies. This understanding will guide your audit approach and help you prioritize areas of focus. Burp Suite's proxy functionality is a cornerstone for auditing cloud security.

You can configure Burp Suite as an intercepting proxy to inspect and modify traffic between your cloud applications and the internet. This allows you to monitor requests and responses, analyze data flows, and identify potential security issues. Before conducting a cloud security audit with Burp Suite, it's crucial to set up the tool correctly. Configure your browser to use Burp Suite as a proxy, and ensure that SSL/TLS interception is properly set up to inspect encrypted traffic. Remember that intercepting SSL/TLS traffic requires installing Burp's CA certificate on your system and browser. Once Burp Suite is properly configured, begin by conducting a thorough reconnaissance of your cloud assets. This includes mapping out the cloud services, endpoints, and APIs that your applications rely on. Use Burp Suite's proxy history and site map features to document these findings. While auditing cloud security, pay particular attention to API endpoints and their authentication mechanisms. Many cloud-based applications interact with APIs to perform various functions. Use Burp Suite's intruder tool to perform automated testing of these endpoints with payloads designed to uncover vulnerabilities. For example, test for SQL injection, XML injection, or other common API vulnerabilities that may exist in your cloud infrastructure. Burp Suite's scanner tool is another valuable asset for cloud security audits. It can automatically scan your cloud applications for a wide range of vulnerabilities, such as cross-site scripting (XSS), SQL injection, and more. However, it's important to configure the scanner to suit your cloud environment's specific needs and authentication methods. Customize scan configurations to include or exclude certain areas of your application, depending on your audit objectives. Burp Suite also provides advanced scanning options for tuning the tool's behavior and accuracy. In cloud security audits, it's common to encounter issues related to misconfigured cloud storage buckets or access control policies. Burp Suite can help identify such misconfigurations by crawling your cloud application and analyzing responses for exposed resources. Pay attention to sensitive data exposure in cloud storage, as unintentional exposure of confidential information can have severe consequences. Burp Suite's content discovery and wordlist features can assist in uncovering hidden or publicly accessible cloud storage objects. As cloud environments often involve complex interactions between different services and components, Burp Suite's collaboration features come in handy. You can share your findings, notes, and audit progress with team members to foster collaboration and streamline the audit process. Furthermore, Burp Suite's extensibility through custom scripts and plugins allows you to tailor the tool to your specific cloud security audit needs. You can develop custom checks for cloud-specific vulnerabilities or create automation scripts to facilitate repetitive tasks. In summary, leveraging Burp Suite for cloud security audits is a practical and efficient approach to identify and address security risks in your cloud environment. By configuring Burp Suite as an intercepting proxy, conducting thorough reconnaissance, testing API endpoints, using the scanner tool, and customizing scans, you can comprehensively *assess the security of your cloud infrastructure. Remember to pay attention to misconfigurations, sensitive data exposure, and collaboration features to enhance the effectiveness of your cloud security audit. Additionally, leverage Burp Suite's extensibility to tailor the tool to your specific cloud environment and audit requirements. Ultimately, a well-executed cloud security audit using Burp Suite can help your organization protect its assets and data in the cloud and maintain a strong security posture.*

Chapter 6: Advanced Reconnaissance Techniques

Understanding the importance of reconnaissance in penetration testing is crucial. Reconnaissance, often referred to as recon, is the process of gathering information about a target. In the context of penetration testing, reconnaissance plays a pivotal role in planning and executing successful assessments. It involves collecting data about the target's infrastructure, systems, applications, and potential vulnerabilities. Reconnaissance can be broadly categorized into two main strategies: passive and active. Passive reconnaissance, as the name suggests, involves collecting information without directly engaging with the target. It's akin to observing and documenting publicly available data. This data includes information such as domain names, IP addresses, subdomains, email addresses, and more. Passive reconnaissance techniques do not send any traffic or requests to the target, making them stealthy and non-intrusive. One common method of passive reconnaissance is using open-source intelligence (OSINT) sources. OSINT involves searching through publicly accessible data on the internet to gather information about the target. This can include searching on search engines, social media platforms, public databases, and even forums. The goal of passive reconnaissance is to build a preliminary profile of the target without alerting or affecting it in any way. It provides valuable insights into the target's online presence and potential attack vectors. Active reconnaissance, on the other hand, is a more direct approach. It involves actively probing the target's systems and infrastructure to gather information. Active reconnaissance techniques interact with the target in some way and may leave traces that the target's security systems can detect. One common method of active reconnaissance is using port scanning. Port scanning involves sending packets to the target's IP addresses to determine which ports are open and what services are running on them. Another active reconnaissance technique is banner grabbing, where an attacker interacts with a service to gather information about its version and configuration. These techniques can help identify potential vulnerabilities and weaknesses in the target's systems. However, it's important to note that active reconnaissance can be more easily detected by intrusion detection systems (IDS) and can potentially trigger alarms. The choice between passive and active reconnaissance depends on the penetration tester's goals and the level of stealth required for the assessment. In many cases, a combination of both passive and active reconnaissance techniques is used to gather comprehensive information about the target. Additionally, it's crucial to stay within the legal and ethical boundaries when conducting reconnaissance activities. Penetration testers should always obtain proper authorization from the target organization before starting any assessment. Once authorization is granted, the reconnaissance phase serves as the foundation for the rest of the penetration *test. It helps penetration testers identify potential entry points, vulnerabilities, and areas of focus for further testing. Reconnaissance can also provide insights into the target's security posture and help tailor the testing approach accordingly. For example, if passive reconnaissance reveals that a target organization uses a specific web application framework, the penetration tester can prepare and select appropriate testing tools and techniques for that framework. The reconnaissance phase is not limited to just technical information. It can also include gathering information about the target's personnel, organizational structure, and security policies. Social engineering is another aspect of reconnaissance that involves manipulating individuals within the target organization to divulge sensitive information. This can include techniques like phishing, pretexting, and tailgating. While social engineering is not strictly part of technical reconnaissance, it is an important aspect of penetration testing as it simulates real-world threats that organizations face. Ultimately, reconnaissance is the foundation upon which the entire penetration testing process is built. It provides the necessary information to make informed decisions about the testing approach, tools, and methodologies. Without a thorough reconnaissance phase, a penetration test may miss critical vulnerabilities or misjudge the target's security posture. In summary, both passive and active reconnaissance strategies are essential components of a successful penetration test. Passive reconnaissance gathers information discreetly from publicly available sources, while active* reconnaissance involves direct interactions with the target. A combination of both strategies provides a comprehensive view of the target's infrastructure, systems, and potential vulnerabilities. Reconnaissance is a critical phase in penetration testing, helping penetration testers identify entry points, vulnerabilities, and areas of focus for further assessment. It also encompasses non-technical aspects such as social engineering, which simulates real-world threats. Overall, reconnaissance is the cornerstone of effective and ethical penetration testing, guiding the entire assessment process.

In the world of cybersecurity and ethical hacking, gathering valuable intelligence is a fundamental aspect of conducting targeted attacks. This intelligence gathering process, often referred to as reconnaissance or recon, forms the initial phase of a comprehensive penetration test. The primary goal of reconnaissance is to collect as much information as possible about the target organization, systems, and potential vulnerabilities. By doing so, penetration testers can assess the security posture of the target, identify potential weaknesses, and plan their attack strategy accordingly. Reconnaissance involves a careful and systematic examination of various aspects of the target, and it plays a pivotal role in the success of the overall assessment. The importance of reconnaissance cannot be overstated, as it provides the foundation upon which the entire penetration test is built. During this phase, penetration testers act as ethical hackers, using various techniques and tools to mimic the activities of malicious attackers. However, it's essential to emphasize that reconnaissance should always be conducted within the bounds of legal and ethical standards, with proper authorization from the target organization. One of the critical distinctions in the reconnaissance phase is the differentiation between passive and active reconnaissance. Passive reconnaissance involves gathering information without directly interacting with the target organization's systems. It's like observing from a distance, relying on publicly available information, and making no attempts to directly communicate with the target. In contrast, active reconnaissance involves direct interactions with the target, such as sending requests, probing for open ports, or attempting to gather information through direct communication with the target's systems. Each type of reconnaissance has its merits and drawbacks, and penetration testers often employ a combination of both to gain a comprehensive view of the target environment. Passive reconnaissance is stealthy and less likely to trigger security alerts on the target's side. It typically involves techniques like open-source intelligence (OSINT) gathering, where publicly accessible data sources are scoured for information. This can include searching for domain names, IP addresses, email addresses, and even employee names and titles on public websites, social media, and online forums. Passive reconnaissance aims to build an initial profile of the target without alerting or affecting it in any way. It provides a starting point for further investigation and allows penetration testers to identify potential attack vectors. Active reconnaissance, on the other hand, involves actions that interact directly with the target environment. Port scanning is a common active reconnaissance technique, where penetration testers send network packets to the target's IP addresses to determine which ports are open and what services are running. Banner grabbing is another active technique, where testers interact with services to gather information about their versions and configurations. While active reconnaissance provides more detailed information about the target, it also carries a higher risk of detection. Intrusion detection systems (IDS) and security monitoring mechanisms are more likely to detect and log these activities. The choice between passive and active reconnaissance often depends on the penetration tester's goals, the level of stealth required, and the specific characteristics of the target environment. In many cases, a balanced approach that combines both passive and active techniques is the most effective way to gather comprehensive intelligence. Reconnaissance goes beyond just technical data; it encompasses a wide range of information, including organizational details, personnel information, and security policies. Social engineering is another crucial aspect of reconnaissance, where ethical hackers simulate real-world attacks that target individuals within the organization. Phishing, pretexting, and tailgating are examples of social engineering techniques used to manipulate individuals into revealing sensitive information. These techniques, while not strictly technical, are essential to consider as they reflect real threats that organizations face. The reconnaissance phase serves as the cornerstone of the penetration testing process, setting the stage for subsequent phases such as scanning, exploitation, and reporting. It helps penetration testers identify potential entry points, vulnerabilities, and areas of focus for further testing. For example, if reconnaissance reveals that a target organization heavily relies on a specific web application framework, penetration testers can prepare and select appropriate testing tools and methodologies tailored to that framework. Additionally, understanding the target's personnel structure and security policies can inform the testing approach, helping penetration testers simulate real-world scenarios more effectively. The reconnaissance phase also helps ethical hackers develop a clearer understanding of the attack surface, which includes the external and internal components of the target's infrastructure. External reconnaissance focuses on information accessible from outside the target organization, while internal reconnaissance dives deeper into the internal network and systems. External reconnaissance may include DNS enumeration to discover subdomains, WHOIS lookup to obtain domain registration information, and enumeration of public-facing services. Internal reconnaissance, on the other hand, might involve network scanning to identify live hosts and services, discovering internal domain names, and mapping the internal network topology. By conducting thorough reconnaissance, penetration testers can prioritize their efforts, streamline the testing process, and maximize the effectiveness of the *assessment. In summary, gathering valuable intelligence through reconnaissance is a critical step in conducting targeted attacks as part of a penetration test. This phase involves a systematic examination of various aspects of the target organization, including technical data, personnel information, and security policies. Reconnaissance provides ethical hackers with the necessary foundation to assess the security posture of the target, identify vulnerabilities, and plan their attack strategy. Differentiating between passive and active reconnaissance techniques is essential, as each has its strengths and limitations. A balanced approach that combines both passive and active techniques is often the most effective way to collect comprehensive intelligence while minimizing the risk of detection. Social engineering techniques should also be considered, as they simulate real-world threats and help penetration testers assess an organization's susceptibility to such attacks. Ultimately, reconnaissance sets the stage for the entire penetration testing process, guiding ethical hackers in making informed decisions and delivering valuable insights to the target organization.*

Chapter 7: Exploiting Network Vulnerabilities

In the realm of cybersecurity and penetration testing, identifying and exploiting network weaknesses is a fundamental skill that ethical hackers must develop. These weaknesses, often referred to as vulnerabilities, are the chinks in the armor of an organization's network infrastructure that malicious attackers could potentially exploit. The goal of ethical hackers is to discover these vulnerabilities before cybercriminals do and help organizations patch them to enhance their security. Identifying network weaknesses begins with a thorough assessment of the network's architecture, components, and configurations. Ethical hackers need to gain a comprehensive understanding of the network's layout, including its external-facing assets and internal components. This assessment typically involves network mapping and enumeration, where hackers discover live hosts, open ports, and running services. Once the network's footprint is established, penetration testers can use various scanning techniques to identify potential weaknesses. Network scanning involves sending packets to target hosts and analyzing their responses to determine which services are running and what vulnerabilities may exist. Common scanning tools like Nmap and Nessus can automate this process, making it efficient and effective. During scanning, ethical hackers may discover open ports associated with well-known services like HTTP, FTP, SSH, and more. Each open port represents a potential entry point into the network, and ethical hackers must investigate these ports further to determine if they are vulnerable to exploitation. One crucial aspect of network scanning is version detection, where hackers determine the specific software and its version running on open ports. This information is invaluable because it allows penetration testers to search for known vulnerabilities associated with the identified software versions. The Common Vulnerabilities and Exposures (CVE) database is an essential resource for ethical hackers, providing a comprehensive list of known vulnerabilities and their associated identifiers. By cross-referencing the discovered software versions with the CVE database, ethical hackers can pinpoint potential weaknesses that require further investigation. In addition to scanning for known vulnerabilities, ethical hackers also look for misconfigurations within the network. Misconfigurations are common culprits that can lead to security weaknesses. For example, a misconfigured firewall rule might inadvertently allow unauthorized access to a critical server, or a weak password policy might expose user accounts to brute-force attacks. Network scanning tools can help identify these misconfigurations by flagging unusual configurations or deviations from best practices. After identifying potential network weaknesses, ethical hackers move on to the exploitation phase, where they attempt to leverage these vulnerabilities to gain unauthorized access or control. It's important to note that exploitation should always be performed with proper authorization and in a controlled environment to avoid causing harm to the target organization. Exploitation often involves using known exploits or creating custom ones to take advantage of the identified vulnerabilities. For example, if a web server is running an outdated version of software known to have a critical security flaw, ethical hackers can attempt to exploit that vulnerability to gain access to the server. During this phase, hackers carefully craft their exploits to maximize the chances of success while minimizing the risk of detection. Exploitation requires a deep understanding of the target system's architecture and the specific vulnerabilities being targeted. Once access is gained, ethical hackers proceed to conduct further assessments, such as privilege escalation and lateral movement, to expand their control within the network. Privilege escalation involves attempting to gain higher levels of access or administrative privileges on compromised systems. Lateral movement focuses on moving laterally across the network to explore other systems and potentially discover additional weaknesses. The ultimate goal of this phase is to provide organizations with a realistic understanding of the potential impact of a successful attack and to help them remediate the identified vulnerabilities. It's crucial to highlight that the exploitation phase should be carried out responsibly and ethically, with a clear focus on security improvement rather than causing harm. Identifying and exploiting network weaknesses is a critical component of penetration testing, helping organizations proactively address vulnerabilities before they can be exploited by malicious actors. This process involves network mapping, enumeration, scanning for vulnerabilities, and exploiting those vulnerabilities responsibly with proper authorization. Ethical hackers rely on tools, databases, and their expertise to identify and assess potential weaknesses effectively. By conducting thorough assessments, ethical hackers provide organizations with valuable insights into their security posture and empower them to enhance their defenses. Post-exploitation techniques play a crucial role in network penetration testing, allowing ethical hackers to maintain access and gather valuable information once they have successfully exploited a target system. These techniques are essential for simulating real-world cyberattacks and helping organizations understand the potential consequences of a security breach. In the context of network penetration testing, post-exploitation is the phase that comes after the initial compromise of a target system. Once an ethical hacker gains access to a network or system, their primary objective shifts from achieving the initial compromise to maintaining a foothold and expanding their control. This shift in focus enables them to mimic the actions of malicious attackers who seek to persistently exploit and navigate within a compromised network. One of the fundamental post-exploitation techniques is privilege escalation, which involves elevating the level of access obtained during the initial compromise. Privilege escalation can take various forms, depending on the target system and the attacker's goals. For example, an ethical hacker may seek to escalate their privileges from *a standard user account to an administrator or root-level access. This enhanced access allows the attacker to perform more advanced actions, such as installing backdoors, exfiltrating sensitive data, or modifying system configurations. Privilege escalation often leverages vulnerabilities or misconfigurations in the target system to grant the attacker higher privileges than initially obtained. Common techniques for privilege escalation include exploiting unpatched software vulnerabilities, taking advantage of weak or default credentials, and abusing insecure configurations. Once privilege escalation is achieved, ethical hackers can effectively control the compromised system and move closer to their goals. Another vital post-exploitation technique is lateral movement, which involves traversing the network to explore and compromise additional systems. Lateral movement mimics the tactics used by attackers to spread within an organization's network once they've breached the perimeter. Ethical hackers use lateral movement to uncover more assets and gather valuable intelligence about the network's architecture and security posture. Lateral movement techniques vary depending on the network's structure and the systems already compromised. Common methods include using stolen credentials to access other systems, exploiting known vulnerabilities, and leveraging built-in Windows tools like PowerShell and Windows Management Instrumentation (WMI) for remote administration. Once lateral movement is successful, ethical hackers can access additional systems, expanding their control and potentially discovering more vulnerabilities. Persistence is another critical post-exploitation aspect, as attackers aim to maintain access to compromised systems over extended periods. Ethical hackers employ various persistence techniques to ensure they can return to the compromised system even if it is rebooted or undergoes security updates. One common method is to create backdoors or hidden user accounts that provide a secret entry point for future access. These backdoors often involve modifying system configurations, adding rogue user accounts, or installing malware that automatically reconnects to the attacker's command and control server. Persistence techniques vary in complexity, but their goal is consistent: to ensure continued access and control over compromised systems. Data exfiltration is another significant post-exploitation activity, where ethical hackers retrieve sensitive information from the compromised network. The objective here is to mimic the actions of cybercriminals who steal valuable data for malicious purposes, such as selling it on the dark web or using it for extortion. Data exfiltration techniques include copying files, capturing network traffic, and using covert channels to transmit data outside the compromised network. Ethical hackers focus on identifying and extracting sensitive information that could have severe consequences if it falls into the wrong hands, such as personally identifiable information (PII) or intellectual property. Furthermore, ethical hackers may engage in lateral movement to compromise additional systems within the network, increasing their access and control. This process continues as ethical hackers explore the network's architecture and gather valuable intelligence, mimicking the actions of malicious attackers. To maintain access to compromised systems, ethical hackers employ persistence techniques that ensure their presence is not easily removed, even in the face of system reboots or updates. Data exfiltration is another critical post-exploitation activity, enabling ethical hackers to retrieve sensitive information and demonstrate the potential impact of a security breach. By simulating these post-exploitation techniques during network penetration testing, ethical hackers help organizations identify and remediate vulnerabilities, ultimately enhancing their overall cybersecurity posture. These techniques provide valuable insights into the potential consequences of a successful attack and enable organizations to strengthen their defenses against real-world threats. In summary, post-exploitation techniques are a crucial aspect of network penetration testing, allowing ethical hackers to maintain access, gather intelligence, and simulate the actions of malicious attackers within a compromised network.*

Chapter 8: Mobile App Exploitation and Reverse Engineering

In the ever-evolving landscape of mobile app security, understanding and mitigating vulnerabilities is of paramount importance to protect sensitive data and user privacy. Advanced mobile app vulnerability exploitation techniques are essential knowledge for security professionals and ethical hackers tasked with assessing the security of mobile applications. As mobile apps continue to gain popularity and become integral parts of our daily lives, they also become attractive targets for cybercriminals seeking to exploit vulnerabilities for malicious purposes. To effectively defend against these threats, it's essential to comprehend the intricacies of mobile app security and the advanced techniques used by attackers. One of the most prevalent and critical vulnerabilities in mobile apps is the OWASP Top Ten Mobile Risks, a list compiled by the Open Web Application Security Project (OWASP) to highlight common security issues. Among these risks, insecure data storage, broken authentication, and insecure communication are significant concerns, and exploiting them can have severe consequences. Insecure data storage, for instance, can lead to unauthorized access to sensitive information stored on the device, such as user credentials, personal data, or financial information. Advanced exploitation techniques may involve reverse engineering the app to access the locally *stored data or leveraging privilege escalation vulnerabilities to gain access to protected storage areas. Broken authentication is another critical concern, as it can lead to unauthorized access to user accounts and sensitive functionality within the app. Advanced attackers may exploit this vulnerability by bypassing authentication mechanisms or intercepting and manipulating authentication tokens. Insecure communication, often associated with the improper use of encryption, can expose sensitive data to eavesdropping attacks. Exploiting this vulnerability may involve conducting man-in-the-middle (MITM) attacks to intercept and decrypt data transmitted between the app and its backend servers. Additionally, advanced attackers may employ traffic analysis techniques to gather intelligence about app users and their behavior. Another mobile app vulnerability that requires advanced exploitation knowledge is insufficient session management, which can lead to unauthorized access and session hijacking. Exploiting this vulnerability may involve analyzing session tokens, session fixation attacks, or session prediction techniques. In some cases, attackers may use reverse engineering or debugging tools to gain insights into the app's session management logic. Mobile app reverse engineering is a crucial skill for advanced vulnerability exploitation, allowing security professionals to analyze an app's code and behavior. Reverse engineering techniques can reveal vulnerabilities, identify insecure data storage, and uncover weaknesses in authentication and session management. Additionally, reverse engineering aids in* understanding the app's communication with backend services, which is vital for identifying and exploiting insecure communication channels. In some cases, attackers may use dynamic analysis tools and emulators to simulate the app's execution environment and monitor its behavior in a controlled environment. Advanced mobile app vulnerability exploitation also encompasses the identification and exploitation of code-level vulnerabilities. For example, attackers may look for insecure code practices such as buffer overflows, injection attacks, or insecure deserialization. Exploiting these vulnerabilities requires a deep understanding of the app's codebase and the ability to craft malicious input to trigger the vulnerability. Another advanced technique is the use of runtime analysis tools to identify and exploit runtime vulnerabilities, such as memory corruption issues or privilege escalation vulnerabilities. These tools help identify weaknesses that may not be apparent through static code analysis alone. Moreover, advanced mobile app vulnerability exploitation includes the use of advanced attack vectors, such as zero-day vulnerabilities. Zero-day vulnerabilities are previously unknown and unpatched security flaws that can be exploited by attackers for maximum impact. Ethical hackers with advanced knowledge may discover and responsibly disclose zero-day vulnerabilities to app developers or security organizations, contributing to the overall improvement of mobile app security. To defend against advanced mobile app vulnerability exploitation, organizations must prioritize security throughout the app development lifecycle. This includes conducting *comprehensive security assessments, code reviews, and penetration testing to identify and remediate vulnerabilities before they can be exploited. Additionally, developers should follow secure coding practices, implement proper authentication and session management, and use encryption and secure communication protocols. Security awareness and training for development teams are crucial to building a strong security foundation for mobile apps. In summary, advanced mobile app vulnerability exploitation techniques are essential knowledge for security professionals and ethical hackers working to assess and enhance the security of mobile applications. Understanding the intricacies of mobile app vulnerabilities, reverse engineering, code-level vulnerabilities, and advanced attack vectors is vital to protecting sensitive data and user privacy in an increasingly mobile-centric world.*

Chapter 9: Cloud Security Assessment and Hardening

In the realm of cloud security, assessing the architecture and configuration of cloud environments is a crucial task for organizations aiming to safeguard their data and applications. Cloud computing has become an integral part of modern IT infrastructure, offering scalability, flexibility, and cost-efficiency. However, these advantages come with a responsibility to ensure that the cloud environment is configured securely and aligned with the organization's security policies. Assessing cloud architecture involves understanding the fundamental design of the cloud infrastructure, which can vary between different cloud service providers, such as Amazon Web Services (AWS), Microsoft Azure, and Google Cloud Platform (GCP). Each cloud provider offers a range of services, including computing, storage, networking, and security, and organizations must design their cloud architecture to meet their specific needs while adhering to best practices in security. One key aspect of assessing cloud architecture is evaluating the network topology, which determines how different cloud resources are interconnected and how data flows within the cloud environment. This assessment ensures that network segments are appropriately isolated, and access controls are in place to prevent unauthorized access to sensitive resources. Additionally, organizations must consider the geographic distribution of their cloud resources to meet data residency and compliance requirements. Another critical element of cloud architecture assessment is the evaluation of identity and access management (IAM) policies and practices. IAM plays a central role in controlling who can access cloud resources and what actions they can perform. Assessing IAM involves reviewing user accounts, permissions, and role assignments to ensure that the principle of least privilege is followed, limiting access to only what is necessary for each user or service. Furthermore, assessing cloud architecture encompasses the examination of data storage and encryption practices. Data security is paramount, and organizations must assess how data is stored, whether it is encrypted at rest and in transit, and whether robust key management practices are in place. This assessment helps identify potential vulnerabilities in data handling and storage, ensuring that sensitive information remains protected. When it comes to cloud configuration assessment, it involves reviewing the settings and configurations of various cloud services and resources. These configurations can significantly impact the security posture of the cloud environment. Common configuration issues include misconfigured access controls, publicly accessible storage buckets, and unsecured database instances. Assessing cloud configuration helps identify and remediate these issues to reduce the attack surface and enhance security. Additionally, organizations should evaluate security groups, network ACLs (Access Control Lists), and firewall rules to ensure that traffic is appropriately restricted and filtered. Cloud providers often offer security tools and services, such as AWS Config, Azure Security Center, and Google Cloud Security Command Center, to help organizations assess and monitor their cloud configurations for compliance and security. Cloud security posture management (CSPM) solutions can also automate the assessment of cloud configuration settings and provide real-time visibility into potential security risks. Assessing cloud architecture and configuration extends to the review of logging and monitoring practices. Effective monitoring is essential for detecting and responding to security incidents and anomalies in the cloud environment. Organizations should assess their logging configurations to ensure that relevant events and activities are logged, and log data is retained for an appropriate duration. Furthermore, they should evaluate their alerting mechanisms to promptly notify security teams of potential threats. Continuous security monitoring and analysis of log data help organizations maintain a proactive security posture. Another aspect of cloud assessment involves compliance and regulatory considerations. Different industries and regions have specific compliance requirements, and organizations must ensure that their cloud architecture and configurations align with these mandates. Assessment should involve auditing cloud configurations against relevant compliance standards, such as the General Data Protection Regulation (GDPR), the Health Insurance Portability and Accountability Act (HIPAA), or the Payment Card Industry Data Security Standard (PCI DSS). Finally, organizations must also consider the shared responsibility model when assessing cloud security. Cloud providers typically share security *responsibilities with their customers, with the provider securing the underlying infrastructure while customers are responsible for securing their data, applications, and configurations. Assessing cloud architecture and configuration necessitates a clear understanding of these shared responsibilities to ensure that security measures are appropriately implemented on both sides. In summary, assessing cloud architecture and configuration is a multifaceted process that involves evaluating the design, settings, and practices within a cloud environment. Organizations must pay careful attention to network topology, IAM policies, data security, configuration settings, monitoring, compliance, and the shared responsibility model to maintain a secure and resilient cloud infrastructure. By conducting thorough assessments and taking corrective actions, organizations can mitigate security risks and leverage the benefits of cloud computing with confidence. In the ever-evolving landscape of cloud computing, implementing security hardening measures is a critical aspect of safeguarding cloud environments. Security hardening refers to the process of securing an information system by reducing its attack surface, minimizing vulnerabilities, and strengthening its overall security posture. When it comes to cloud environments, the need for security hardening is even more pronounced due to the shared responsibility model, where both the cloud service provider and the customer have roles in securing the infrastructure. Next, we will explore the essential security hardening measures that organizations should implement to enhance the security of their cloud environments. One of the fundamental steps in security hardening for cloud environments is identity and access management (IAM). IAM policies and practices play a pivotal role in controlling access to cloud resources and services. Organizations should adopt the principle of least privilege, ensuring that users and services have only the permissions necessary to perform their specific tasks. This minimizes the risk of unauthorized access and reduces the potential for privilege escalation. Furthermore, organizations should regularly review and audit IAM policies to ensure they align with evolving security requirements. Another crucial aspect of security hardening is network security. Securing the network in a cloud environment involves implementing robust firewall rules, security groups, and network ACLs to control traffic flow and restrict access to trusted sources only. Additionally, organizations should consider implementing a virtual private cloud (VPC) or virtual network to isolate different parts of their cloud infrastructure. By segmenting the network, organizations can limit lateral movement in the event of a breach. Hardening the network also involves monitoring network traffic for suspicious activities, which can be achieved using intrusion detection and prevention systems (IDPS) and network monitoring tools. Next, data security is paramount when it comes to security hardening in the cloud. Encrypting data both at rest and in transit is essential to protect sensitive information from unauthorized access. Cloud providers typically offer encryption services and key management solutions to help organizations secure their data. It's essential to configure these services correctly and manage encryption keys effectively. Additionally, organizations should implement data loss prevention (DLP) measures to prevent the accidental or intentional leakage of sensitive data. Another vital aspect of security hardening is vulnerability management. Regularly scanning cloud environments for vulnerabilities is essential to identify weaknesses that could be exploited by attackers. Organizations should use vulnerability scanning tools to assess their cloud infrastructure, and they should have processes in place to patch or remediate identified vulnerabilities promptly. Security patch management is a crucial part of vulnerability management to ensure that systems are up to date with the latest security patches. Additionally, organizations should consider implementing intrusion detection systems (IDS) and intrusion prevention systems (IPS) to detect and block malicious activities in real time. Authentication mechanisms are also a focal point in security hardening. Implementing multi-factor authentication (MFA) adds an extra layer of security by requiring users to provide multiple forms of verification before gaining access to cloud resources. This significantly enhances account security and reduces the risk of unauthorized access due to stolen or compromised credentials. Organizations should also implement strong password policies and enforce password complexity requirements to thwart password-based attacks. Furthermore, they should consider using single sign-on (SSO) solutions to centralize authentication and improve user access management.*

When it comes to securing cloud assets, asset management is a critical component of security hardening. Organizations must maintain an up-to-date inventory of all cloud resources, including virtual machines, storage buckets, databases, and more. This inventory helps ensure that no resources are inadvertently left exposed or unsecured. Moreover, organizations should implement resource tagging and labeling to categorize and organize assets effectively. Proactive monitoring and alerting are indispensable security hardening measures. Continuous monitoring of cloud environments enables organizations to detect and respond to security incidents promptly. Security information and event management (SIEM) systems can aggregate and analyze log data from various sources, providing valuable insights into potential threats. Implementing robust incident response and recovery plans is essential to minimize damage in case of a security breach. Organizations should regularly test their incident response procedures to ensure that they are effective and can be executed swiftly. Security hardening also extends to container and serverless computing environments. For containers, organizations should follow best practices for securing container images, orchestrators (e.g., Kubernetes), and the underlying infrastructure. Serverless security involves securing the code and functions running in serverless environments, configuring appropriate access controls, and monitoring for suspicious activity. Lastly, compliance plays a significant role in security hardening for cloud environments. Organizations must ensure that their *cloud infrastructure adheres to industry-specific regulations and compliance standards. This may involve implementing specific controls, documenting security policies and procedures, and regularly conducting compliance audits. In summary, implementing security hardening measures is an ongoing process that requires diligence and a comprehensive approach to security. Organizations should prioritize IAM, network security, data security, vulnerability management, authentication, asset management, monitoring, and compliance in their security hardening efforts. By taking a proactive stance on security, organizations can better protect their cloud environments from evolving threats and maintain a robust security posture.*

Chapter 10: Automating Multi-Platform Assessments

Scripting and automation are powerful techniques that can significantly enhance your efficiency and effectiveness when using Burp Suite. By automating repetitive tasks and scripting custom actions, you can streamline your workflow and save valuable time during your penetration testing and security assessments. Next, we'll explore the world of scripting and automation within Burp Suite and how these capabilities can benefit your security testing efforts. Burp Suite provides a robust scripting environment that allows you to create custom scripts using the Python programming language. Python is a widely-used language known for its simplicity and readability, making it an excellent choice for scripting within Burp Suite. With Burp's extensibility, you can write scripts to automate various tasks, from simple to complex, tailored to your specific testing requirements. One of the primary benefits of scripting in Burp Suite is the ability to automate the repetitive aspects of your security testing. For example, you can write scripts to automatically perform common actions like spidering a web application, submitting forms, or interacting with APIs. This automation not only saves time but also reduces the likelihood of human error that can occur when performing these tasks manually. Scripts can also be used to extend Burp's functionality beyond its built-in capabilities. For instance, you can create custom scanning checks or add support for proprietary web applications or APIs. This flexibility allows you to adapt Burp Suite to meet the unique challenges of your security assessments. Another advantage of scripting in Burp Suite is the ability to create custom reporting and data analysis tools. You can develop scripts that generate customized reports, extract and analyze specific data from scan results, or integrate Burp Suite with other security tools and systems. This level of customization enables you to tailor your reporting to the needs of your organization or clients. While scripting in Burp Suite offers immense flexibility, it's essential to approach it with some best practices in mind. First, ensure that your scripts are well-documented to make them understandable and maintainable over time. Clear and concise comments within your code can be invaluable for both yourself and others who may work with your scripts. Additionally, consider error handling in your scripts to gracefully handle unexpected situations and prevent crashes. Error handling can help your scripts continue functioning even when they encounter issues during execution. Furthermore, it's crucial to thoroughly test your scripts in a safe environment before using them in a production or testing environment. Testing helps identify and resolve any issues or bugs in your scripts, ensuring that they perform as expected during your assessments. Burp Suite provides a dedicated interface for managing and running your scripts, known as the "Extender" tab. Here, you can write and edit your scripts, load and unload script extensions, and view script output and errors. The "BApp Store" within Burp Suite offers a repository of *pre-made extensions and scripts that you can install and use, saving you time and effort. When developing custom scripts, you can leverage Burp's extensive API (Application Programming Interface) to interact with various components of the tool. This API exposes functions and objects that allow your scripts to access and manipulate HTTP messages, scan issues, configuration settings, and more. By understanding and utilizing the API, you can harness the full power of Burp Suite in your scripting endeavors. Beyond Python scripting, Burp Suite also supports other scripting languages such as JavaScript, Ruby, and Groovy. You can choose the scripting language that you are most comfortable with or that best suits your specific needs. This flexibility allows you to work with the language you are most proficient in, making the scripting process more accessible and efficient. In addition to scripting, Burp Suite offers a feature known as "Session Handling Rules," which allows you to automate session management and adapt your testing to various authentication mechanisms. These rules enable you to define conditions and actions that Burp Suite should take when certain criteria are met during a testing session. For example, you can use session handling rules to automatically log in to an application, handle session tokens, or perform actions based on specific HTTP responses. This feature simplifies the process of dealing with complex authentication scenarios and helps ensure that your testing remains comprehensive. Another valuable automation feature in Burp Suite is its support for macros. A macro is a series of HTTP requests and responses that represent a sequence of interactions with a web application. You can record these interactions in Burp Suite and then replay them as a single action. Macros are particularly useful for automating multi-step processes within web applications, such as navigating through a workflow or submitting a series of forms. By recording and replaying macros, you can expedite your testing and ensure that all required steps are covered. To conclude, scripting and automation are indispensable tools in the arsenal of any Burp Suite user. By harnessing the power of scripting, you can automate tasks, extend Burp's functionality, customize reporting, and adapt to the unique challenges of your security assessments. Whether you're a novice or an experienced user, incorporating scripting and automation into your workflow can enhance your productivity and help you uncover vulnerabilities more effectively. As you delve into the world of Burp Suite scripting, remember to follow best practices, test your scripts thoroughly, and leverage the extensive API and features provided by Burp Suite to maximize your testing capabilities.*

**BOOK 4
BURP SUITE NINJA
ADVANCED STRATEGIES FOR ETHICAL HACKING AND
SECURITY AUDITING**

ROB BOTWRIGHT

Chapter 1: The Path to Burp Suite Mastery

Setting clear and achievable learning goals is a fundamental step in any educational journey, including your exploration of Burp Suite and web application security. When embarking on this endeavor, it's essential to define what you hope to accomplish and establish a roadmap to guide your efforts. Learning goals serve as a compass, helping you stay focused and motivated throughout your studies and practical experiences. The first step in setting your learning goals is to identify your specific areas of interest within web application security and Burp Suite. Perhaps you're drawn to vulnerability scanning, penetration testing, or the intricacies of web protocols. Or maybe you're eager to master the automation capabilities of Burp Suite or explore advanced security testing techniques. Whatever your interests, recognizing them is the initial stage in crafting meaningful learning objectives. Once you've pinpointed your interests, the next step is to break down your goals into smaller, manageable components. Consider what skills, knowledge, or competencies you need to develop within your chosen area of focus. For example, if your goal is to become proficient in web application scanning with Burp Suite, you might need to acquire expertise in identifying vulnerabilities like SQL injection, cross-site scripting (XSS), or security misconfigurations. You may also want to learn how to configure and customize Burp Suite to suit your testing needs effectively. These sub-goals will help you structure your learning journey and monitor your progress along the way. As you set your learning goals, it's important to make them SMART: Specific, Measurable, Achievable, Relevant, and Time-bound. Specific goals clearly define what you want to achieve, leaving no room for ambiguity. Instead of a vague objective like "I want to get better at web security," you might state, "I aim to master SQL injection testing with Burp Suite." Measurable goals allow you to track your progress and determine when you've achieved them. You can measure your proficiency in a particular skill, such as identifying XSS vulnerabilities, by quantifying the number of successful identifications in your testing. Achievable goals are realistic and attainable within your capabilities and available resources. Setting an unattainable goal can lead to frustration and demotivation. For example, aiming to become a web security expert overnight is not achievable, but becoming proficient in a specific aspect of it within a few months might be. Relevant goals align with your interests, values, and career aspirations. They should have a clear connection to your overall objectives, ensuring that your efforts are purposeful and meaningful. If your ultimate career goal is to become a penetration tester, then goals related to mastering Burp Suite and web application security are highly relevant. Finally, time-bound goals have a set timeframe for completion. This helps create a sense of urgency and prevents procrastination. You might set a goal to become proficient in a particular skill or aspect of Burp

Suite within six months. Once you've established SMART learning goals, consider creating a visual roadmap or plan to outline the steps needed to achieve them. This roadmap can serve as a visual reminder of your objectives and provide a clear path forward. For example, if your goal is to excel in web application authentication testing, your roadmap might include steps such as studying authentication mechanisms, practicing with Burp Suite's tools, and conducting real-world assessments. Breaking these steps into smaller, sequential tasks can make the journey more manageable. Furthermore, it's beneficial to seek out additional resources and learning materials that align with your goals. These can include online courses, tutorials, books, documentation, and communities where you can engage with experts and fellow learners. For instance, if you're passionate about mastering Burp Suite's scripting capabilities, you might explore Python programming courses or join forums dedicated to Burp Suite scripting discussions. Learning from others and sharing experiences can enrich your understanding and provide valuable insights. Remember that setting and achieving learning goals is an ongoing process. It's natural to encounter challenges and setbacks along the way, but these should be viewed as opportunities for growth. Regularly review and adjust your goals as needed to accommodate changes in your interests or circumstances. Additionally, celebrate your successes, no matter how small they may seem. Recognizing your achievements can boost your motivation and reinforce your commitment to your learning journey. In summary, *setting effective learning goals is a crucial foundation for your exploration of Burp Suite and web application security. By defining your areas of interest, breaking down objectives into SMART goals, creating a roadmap, seeking relevant resources, and remaining adaptable, you'll be well-prepared to embark on a rewarding and educational adventure. With determination, patience, and a clear sense of purpose, you can make significant strides in your knowledge and expertise in web application security and Burp Suite. Exploring resources for mastering Burp Suite and web application security is an exciting and essential aspect of your learning journey. To become proficient in this field, you'll want to leverage a variety of educational materials and tools that cater to different learning styles and preferences. One of the most accessible resources available is the Burp Suite documentation, which serves as your comprehensive guide to understanding the tool's features and capabilities. The documentation is well-organized, featuring detailed explanations, examples, and usage scenarios for each component of Burp Suite. Whether you're a beginner seeking basic guidance or an advanced user delving into the intricacies of the tool, the documentation can be an invaluable reference. Online courses and tutorials are another valuable resource to consider. Several platforms offer courses specifically designed to teach Burp Suite and web application security. These courses often come with video lessons, hands-on labs, quizzes, and assignments, providing a structured and engaging learning experience. Some even offer certification upon completion, which can enhance your credibility in the field. Books dedicated to Burp Suite and web application security are also abundant. Authors with expertise in the field have written extensively on the subject, offering in-depth insights and practical knowledge. These books cover a wide range of topics, from the fundamentals of web application security to advanced techniques for using Burp Suite effectively. Consider exploring titles like "Web Application Security Testing with Burp Suite" or "Mastering Burp Suite Community Edition" to expand your knowledge. Interactive labs and platforms provide hands-on practice, allowing you to apply your skills in a controlled environment. These labs often simulate real-world scenarios, enabling you to test your abilities in identifying and exploiting vulnerabilities. Platforms like PortSwigger Web Security Academy offer a collection of free labs that align with the OWASP Top Ten, providing an excellent opportunity for skill development. Online forums and communities can be invaluable for networking and seeking guidance from peers and experts. Platforms like Stack Overflow, Reddit's /r/netsec, and Burp Suite's official community forum are excellent places to ask questions, share experiences, and gain insights from others in the field. Participating in discussions and actively engaging with these communities can help you stay updated on the latest trends and challenges in web application security. Professional organizations and conferences offer opportunities for networking and expanding your knowledge. Organizations like OWASP (Open Web Application Security Project) host events and conferences worldwide, providing a platform to connect with industry professionals and attend informative talks and workshops. Attending these events can help you stay informed about emerging threats, best practices, and the latest developments in the field. Security blogs and podcasts are excellent resources for staying current with the rapidly evolving field of web application security. Prominent security researchers and practitioners often share their insights, experiences, and analysis through these mediums. Subscribing to blogs or podcasts relevant to your interests can provide regular updates and fresh perspectives on the subject. Practical exercises and challenges can solidify your skills by applying them to real-world scenarios. Platforms like Hack The Box, TryHackMe, and PortSwigger's Web Security Academy offer various challenges and labs for honing your web application security skills. These platforms provide a safe and legal environment to practice penetration testing and vulnerability analysis. Mentorship and peer learning can accelerate your growth in web application security. Seeking guidance from experienced professionals or collaborating with peers who share similar goals can provide valuable insights and support. Mentorship programs and local cybersecurity meetups can facilitate these connections. Consider joining or forming study groups to tackle challenges together. Online video content and webinars are engaging resources for visual learners. Many experts in the field create video tutorials, walkthroughs, and webinars that demonstrate practical techniques using Burp Suite and other security tools. These videos can* complement your learning by providing step-by-step guidance and demonstrations. Additionally, Burp Suite's community edition is an excellent starting point for beginners. It offers robust functionality for web security testing, including scanning, crawling, and intercepting requests and responses. While the free version has some limitations compared to the professional edition, it's a powerful tool for learning and practicing web application security. As you progress in your learning journey, you may consider upgrading to the professional version to access advanced features. Lastly, it's essential to stay up to date with the latest security news and developments. Following security-focused websites, newsletters, and Twitter accounts can help you remain informed about emerging threats and vulnerabilities. By being aware of current security trends, you'll be better equipped to address new challenges in the field. In summary, mastering Burp Suite and web application security requires a multifaceted approach that leverages a variety of resources and tools. The Burp Suite documentation, online courses, books, interactive labs, forums, communities, organizations, conferences, blogs, podcasts, practical exercises, mentorship, video content, and staying informed are all valuable components of your learning toolkit. Combining these resources and tailoring your learning journey to your specific goals and preferences will lead to a deeper understanding of web application security and proficiency in using Burp Suite effectively. Remember that learning is an ongoing process, and staying curious and adaptable is key to achieving mastery in this dynamic field.

Chapter 2: Advanced Burp Suite Configuration and Customization

Customizing Burp Suite to suit your specific testing needs is a crucial step in becoming proficient in web application security. Burp Suite is a versatile tool, and tailoring it to your requirements can significantly enhance your efficiency and effectiveness as a penetration tester or security professional. Next, we'll explore the various aspects of customizing Burp Suite and how these customizations can benefit your testing efforts. Let's start by discussing why customization is essential. Every web application is unique in its architecture, technologies used, and potential vulnerabilities. As a tester, you need a tool that adapts to these differences and allows you to target your testing efforts effectively. This is where Burp Suite's flexibility shines. One of the primary ways to customize Burp Suite is by configuring its proxy settings. The proxy is at the core of Burp Suite's functionality, allowing you to intercept, inspect, and manipulate requests and responses between your browser and the target web application. By adjusting the proxy settings, you can define how Burp Suite interacts with the traffic passing through it. For example, you can specify interception rules to capture specific requests or responses based on URLs, parameters, or other criteria. This feature allows you to focus your testing on particular areas of the application. Another essential customization is configuring the scope of your testing. The scope defines which parts of the target application are within the testing boundaries. You can specify the inclusion and exclusion criteria for URLs, hosts, and folders. By carefully defining your scope, you ensure that your testing remains targeted and avoids unintended interactions with external resources. Beyond the scope, Burp Suite allows you to fine-tune its automated scanning and crawling capabilities. You can configure scan policies to control how aggressively Burp Suite scans for vulnerabilities. Customizing scan policies enables you to balance between thoroughness and speed based on the target application's size and complexity. Furthermore, you can configure issue definitions to customize how Burp Suite identifies and categorizes vulnerabilities. These customizations help tailor Burp Suite's reporting and alerting to your organization's specific needs. In addition to automated scanning, manual testing is a significant part of web application security assessments. Burp Suite provides a powerful set of tools for manual testing, and you can customize these tools to streamline your workflow. For example, you can create custom macros to automate sequences of interactions with the application, simplifying repetitive tasks. By utilizing macros, you save time and reduce the risk of human error. Burp Suite's extensibility through extensions is another aspect that allows for extensive customization. Extensions are scripts and plugins that you can integrate into Burp Suite to add new features and capabilities. You can develop your extensions or leverage the many existing ones created by the community. Custom extensions can be tailored to address specific testing requirements or to integrate with other security tools in your workflow. Furthermore, Burp Suite's user interface can be customized to enhance your productivity. You can rearrange panels, create custom tab layouts, and configure hotkeys to access frequently used features quickly. This level of customization allows you to design a workspace that fits your preferred workflow and testing style. When it comes to handling sessions and authentication during testing, Burp Suite offers features to assist with customization. You can configure session handling rules to automatically log in or maintain sessions as you navigate the application. These rules are highly customizable and can adapt to various authentication mechanisms, including form-based, HTTP basic, and token-based authentication. This capability simplifies testing scenarios where you need to maintain a logged-in state throughout your assessment. Moreover, Burp Suite supports customizing its automated scans for specific vulnerabilities and technologies. You can adjust scan checks and insertion points to target vulnerabilities that are more common or relevant to the target application. This level of granularity allows you to focus your testing on areas where you suspect vulnerabilities may exist. In addition to scan checks, Burp Suite's content discovery and bruteforcing capabilities can be customized. You can define custom wordlists, rules, and filters to fine-tune the content discovery process. By doing so, you can improve the efficiency of finding hidden or restricted *resources within the application. Burp Suite's reporting capabilities are highly customizable to ensure that your assessment findings are well-documented and communicated effectively. You can create custom report templates, incorporating your organization's branding and specific details. These templates can be tailored to include or exclude specific sections, such as a detailed technical description of vulnerabilities, proof-of-concept requests, or recommendations for remediation. This customization helps produce clear and actionable reports for stakeholders. Finally, Burp Suite's configuration and customization options extend to performance tuning. When dealing with large-scale applications or extensive testing, optimizing the tool's performance is essential. You can adjust memory settings, thread pools, and connection limits to ensure that Burp Suite operates efficiently without overwhelming your system resources. These performance tweaks are valuable for handling challenging testing scenarios. In summary, customizing Burp Suite is a fundamental aspect of harnessing its full potential as a web application security testing tool. By configuring its proxy settings, defining the scope of your testing, adjusting scanning and crawling policies, and tailoring the user interface to your workflow, you can adapt Burp Suite to meet your specific testing requirements. Leveraging extensions, customizing manual testing tools, handling sessions, and fine-tuning scanning and reporting capabilities further enhance your ability to uncover vulnerabilities effectively. Ultimately, the flexibility and extensive customization options provided by Burp Suite empower you to become a more efficient and skilled web application security tester. The next chapter will delve deeper into the practical aspects of customizing Burp Suite for various testing scenarios.*

Chapter 3: Leveraging Burp Macros and Extensions

Automation plays a pivotal role in web application security testing, and Burp Suite offers robust automation capabilities through the use of macros. Next, we will explore the world of Burp macros, uncovering how they can help you streamline your testing workflow and save valuable time. Burp macros allow you to record a sequence of interactions with a web application and then replay them automatically. Imagine a scenario where you need to perform a series of steps to authenticate, navigate to a specific page, and submit data to various forms within the application. Manually executing these steps can be time-consuming and prone to human error. This is where macros come into play. You can record your interactions with the application once and replay them as many times as needed. Burp Suite captures the HTTP requests and responses during the recording, allowing you to automate complex, multi-step testing scenarios with ease. To start using macros, you'll first need to configure your browser to route traffic through Burp Suite's proxy. Once your browser is configured to use Burp as a proxy, you can initiate a new macro recording session. During the recording, Burp Suite intercepts and logs all the HTTP traffic between your browser and the target application. You can perform actions within the browser, such as clicking links, submitting forms, and interacting with web pages, just as you would during manual testing. Burp Suite faithfully records each HTTP request and response, creating a comprehensive record of your interactions. This recording can include various types of requests, including GET, POST, and AJAX requests. One of the key benefits of using macros is the ability to parameterize your interactions. In other words, you can replace specific values in your recorded requests with placeholders that Burp Suite will substitute with different data during replay. For example, you can replace a username and password with placeholders, allowing you to test different login credentials without manually modifying the requests. Parameterization makes your macros dynamic and adaptable for various testing scenarios. After recording your macro, you can edit it within Burp Suite to fine-tune the interactions or add additional requests as needed. This editing capability is invaluable when you encounter more complex testing scenarios that require customization. Once your macro is ready, you can replay it as many times as needed. This automation greatly accelerates repetitive testing tasks and ensures consistent, reliable testing results. Furthermore, you can configure macros to run with different sets of data, making them highly versatile. Imagine you need to test an application's search functionality with various keywords or test a form with different input combinations. By parameterizing your macros and running them with different data sets, you can efficiently perform comprehensive testing without manual repetition. In addition to running macros manually, Burp Suite provides automation options to schedule and repeat macro executions automatically. You can set up macros to run at specific intervals, allowing you to perform continuous testing or monitor an application for vulnerabilities over time. Automating macros in this way is especially useful for assessing an application's security posture in a real-world, dynamic environment. Moreover, macros can be integrated into Burp Suite's scanning process. By incorporating macros into automated scanning, you can ensure that authenticated areas of an application are tested thoroughly. For example, if you need to scan a web application that requires a login, you can configure Burp Suite to run the login macro before initiating the scan. This ensures that the scanner can access authenticated areas of the application and discover vulnerabilities that may be hidden behind a login mechanism. To further enhance the power of macros, you can combine them with Burp's extensibility through custom scripts and extensions. This allows you to create custom logic and decision-making within your macros. For instance, you can use a custom script to extract a session token from a response and then use that token in subsequent requests, simulating a more realistic user behavior. Custom scripts and extensions extend the automation capabilities of macros, making them adaptable to even the most complex testing scenarios. As you become more proficient in working with Burp macros, you'll discover numerous creative ways to leverage their power. You can use macros for more than just authentication and session management. Consider scenarios where you need to test for business logic flaws, such as testing shopping cart functionality, performing multi-step transactions, or simulating user interactions with rich client-side applications. Burp macros can be customized to handle these complex use cases, providing you with the flexibility and efficiency needed to identify vulnerabilities effectively. To conclude, automating tasks with Burp macros is a valuable skill that can significantly enhance your web application security testing capabilities. Macros allow you to record and replay sequences of interactions with web applications, automating repetitive tasks and ensuring consistent testing results. Their parameterization, editing capabilities, and integration with Burp's scanning process make them versatile and adaptable to various testing scenarios. By combining macros with custom scripts and extensions, you can address complex testing requirements and simulate realistic user behaviors. With practice and creativity, you can leverage Burp macros to become a more efficient and effective web application security tester. In the next chapter, we will delve deeper into Burp Suite's extensibility and explore the world of custom extensions and scripts.

Chapter 4: Mastering Burp's Intricate Scanning Techniques

Next, we'll explore advanced web application scanning strategies that go beyond the basics to help you uncover even the most elusive vulnerabilities. While automated scanning tools like Burp Suite are incredibly powerful, they may not always detect subtle or custom vulnerabilities. To truly master web application security testing, you need to complement automated scanning with advanced manual testing techniques. One such technique is manual source code review, where you examine the application's source code for vulnerabilities that may not be apparent through dynamic scanning alone. By analyzing the code, you can identify issues such as hardcoded secrets, insecure function calls, or custom security mechanisms that require a more nuanced approach to testing. Additionally, manual code review allows you to understand the application's logic and data flow, helping you pinpoint potential vulnerabilities that automated scanners might overlook. Another advanced technique is using specialized fuzzing tools to test for specific vulnerabilities like XML injection or LDAP injection. These tools generate malformed or unexpected input data to trigger potential vulnerabilities, providing a more targeted approach to testing. By tailoring your testing efforts to the application's technology stack and architecture, you can maximize your chances of discovering hidden vulnerabilities. Beyond manual source code review and specialized fuzzing, advanced web application scanning also involves in-depth knowledge of various attack vectors. You should be familiar with the OWASP Top Ten and other common web application vulnerabilities, such as SQL injection, cross-site scripting (XSS), and remote code execution. Understanding these vulnerabilities at a deep level allows you to craft sophisticated attacks that mimic real-world threats. For example, instead of relying solely on automated SQL injection payloads, you can craft custom SQL queries that exploit the application's specific database structure. This level of customization can reveal vulnerabilities that automated tools might miss. In addition to attack vectors, mastering advanced web application scanning requires expertise in bypassing security mechanisms. Many web applications employ security controls like Web Application Firewalls (WAFs) or input validation filters. To effectively test these applications, you need to develop evasion techniques that can bypass such security controls. This may involve crafting payloads that obfuscate attack payloads or splitting attacks into multiple requests to evade detection. In some cases, you might need to explore lesser-known attack techniques, such as Server-Side Template Injection (SSTI) or Business Logic Flaws. These vulnerabilities often require a deep understanding of the application's architecture and logic to identify. To test for them effectively, you may need to reverse engineer client-side code or manipulate application parameters in unique ways. Moreover, advanced web application scanning extends to testing RESTful APIs and GraphQL endpoints. Modern web applications often rely on APIs for data exchange, making them a potential attack surface. Testing APIs requires a different approach than testing traditional web applications. You should become proficient in crafting custom requests, manipulating JSON or XML payloads, and *understanding the intricacies of the application's API endpoints. Advanced techniques like parameter pollution or prototype pollution should also be in your toolkit when testing APIs. When it comes to web application scanning, timing and sequencing of requests are critical. Automated scanners often follow predefined sequences, but real-world attackers don't always adhere to such patterns. As an advanced tester, you should experiment with different timing and sequencing of requests to identify race conditions, concurrency issues, or vulnerabilities triggered by specific sequences of actions. Furthermore, understanding session management mechanisms is crucial for detecting vulnerabilities related to authentication and authorization. You should be adept at manipulating session tokens, cookies, and other session-related parameters to bypass access controls or escalate privileges. In some cases, advanced testing may require crafting custom payloads to exploit vulnerabilities that are unique to the application. For example, if you suspect an XML external entity (XXE) vulnerability, you might create a custom Document Type Definition (DTD) that reveals sensitive information. Similarly, if you encounter a deserialization vulnerability, you might construct a custom payload that triggers remote code execution. The ability to think creatively and adapt your testing approach to the specific application is a hallmark of advanced web application scanning. In summary, advanced web application scanning goes beyond automated tools and requires a deeper understanding of attack vectors, evasion techniques, and the application's technology stack. Manual source code review, specialized fuzzing, and the mastery of various vulnerabilities and security controls are* essential components of advanced testing. Developing custom payloads, understanding session management, and experimenting with different request timing and sequencing are also key skills. By combining these techniques and continuously expanding your knowledge, you can become a highly effective web application security tester who uncovers even the most sophisticated vulnerabilities. In the next chapter, we will delve into the world of penetration testing methodologies and explore how to plan and execute comprehensive tests that assess an application's security posture thoroughly. Next, we will dive deep into the process of fine-tuning Burp Scanner to achieve the highest level of precision and accuracy in your web application security assessments. Burp Scanner is a powerful automated scanning tool that can help you identify vulnerabilities quickly. However, like any automated tool, it requires careful configuration and fine-tuning to ensure it produces reliable results. Fine-tuning Burp Scanner is not a one-size-fits-all process, as the optimal settings can vary depending on the specific web application you are testing. One of the first steps in fine-tuning Burp Scanner is to define the scope of your scan accurately. You should clearly define the target URLs, including which parts of the application you want to scan and which parts you want to exclude. By narrowing down the scope, you reduce unnecessary scan traffic and false positives. Additionally, you can configure Burp Scanner to follow a specific crawl pattern. For example, you can instruct it to prioritize scanning high-value pages, such as authentication or admin panels, before less critical areas of the application. This ensures that Burp Scanner focuses its efforts on areas where vulnerabilities are more likely to be found. Fine-tuning scan settings also involves selecting the appropriate scan checks based on the vulnerabilities you want to detect. Burp Scanner offers various scan check options, including those for SQL injection, cross-site scripting (XSS), and more. By enabling or disabling specific checks, you can tailor the scan to your requirements, reducing noise and improving efficiency. Moreover, you can adjust the payload options for each scan check to test different attack vectors or payloads. For example, you can configure the SQL injection check to use specific SQL payloads that are relevant to the database technology used by the application. Fine-tuning scan speed is another critical aspect of optimizing Burp Scanner. You can adjust the scan speed setting based on the application's size and complexity. Higher scan speeds increase the number of requests per second, which can help complete the scan faster but may also increase the risk of causing disruptions or lockouts. Conversely, lower scan speeds reduce the rate of requests, which is more suitable for applications that are sensitive to scanning traffic. Furthermore, Burp Scanner allows you to configure scan policies, which define how it interacts with the application during the scan. You can specify how it handles authentication, handles session management, and deals with anti-CSRF tokens. Fine-tuning these policies is essential to ensure that Burp Scanner can navigate the application successfully and identify vulnerabilities within authenticated areas. In some cases, you may need to provide custom authentication credentials or implement session handling rules to help Burp Scanner maintain a valid testing state. It's important to monitor the progress of your scans regularly. Burp Suite provides real-time feedback on the scan's status, including the number of requests sent, responses received, and vulnerabilities discovered. By keeping an eye on these metrics, you can quickly identify any issues or anomalies during the scan. If you notice unexpected behavior or scan failures, you can adjust your settings accordingly to address the problem. Additionally, Burp Scanner offers the option to pause or stop a scan at any time, giving you control over the scanning process. Another aspect of fine-tuning Burp Scanner involves managing the handling of errors and exceptions. Web applications can behave unpredictably, and not all errors indicate vulnerabilities. You can configure Burp Scanner to ignore certain types of errors or exceptions to reduce the number of false positives. Additionally, Burp Scanner provides the capability to customize its behavior through scan-specific issue definitions. You can define issue patterns that trigger when certain conditions are met during the scan. For example, you can create a custom issue pattern to identify specific error messages or content that indicates a vulnerability. This allows you to extend the scanner's detection capabilities to cover unique vulnerabilities in your target application. Moreover, Burp Scanner provides advanced options for handling edge cases and challenges. For example, if an application employs CAPTCHA challenges or multi-step authentication processes, you can configure Burp Scanner to solve them automatically or prompt you for manual intervention. Fine-tuning Burp Scanner also involves optimizing its performance by adjusting the number of concurrent scan tasks and threads. You can increase or decrease these settings based on your system's capabilities and the application's responsiveness. Optimizing concurrency can help balance scan efficiency and resource utilization. It's worth noting that Burp Scanner offers the ability to pause and resume scans at any time. This feature is useful when you want to pause a scan temporarily to investigate a potential vulnerability or make adjustments to your scan configuration. Once you've made the necessary changes, you can resume the scan from where it left off. In summary, fine-tuning Burp Scanner is a crucial step in ensuring the accuracy and precision of your web application security assessments. By carefully configuring scan settings, defining scope, monitoring progress, managing errors, and customizing issue definitions, you can optimize the tool's performance and increase its ability to detect vulnerabilities. In the next chapter, we'll explore advanced techniques for manual testing and exploitation of web application vulnerabilities, taking your skills to the next level.

Chapter 5: Exploiting Advanced Web Application Vulnerabilities

Next, we will delve into the art of exploiting SQL injection vulnerabilities like a professional penetration tester. SQL injection is one of the most prevalent and dangerous web application vulnerabilities, and mastering its exploitation is essential for comprehensive security assessments. To become proficient in exploiting SQL injection vulnerabilities, you need to understand not only the technical aspects of the attack but also the underlying principles of SQL databases. SQL, which stands for Structured Query Language, is a powerful language used for managing and querying relational databases. It allows users to interact with databases, retrieve data, insert records, update information, and delete entries. SQL injection occurs when an attacker manipulates an application's input in such a way that it injects malicious SQL queries into the application's database. These queries can potentially execute arbitrary commands on the database, leading to data leakage, data manipulation, or even complete system compromise. The first step in exploiting SQL injection is identifying the vulnerability in the target application. You can begin by testing input fields, such as search boxes, login forms, or any parameter that interacts with a database, for signs of susceptibility. Common indicators include error messages, unusual application behavior, or differences in response times. Once you've *identified a potential SQL injection point, you can start crafting malicious payloads to exploit the vulnerability. SQL injection payloads are essentially SQL statements that you inject into the application's input fields. The goal is to manipulate the query in a way that extracts sensitive data or performs unintended actions on the database. An example of a simple SQL injection payload is appending a single quote (') to input fields and observing if the application generates an error. If it does, it's a strong indication of a potential SQL injection vulnerability. Beyond simple payloads, more sophisticated techniques involve understanding the database's schema and crafting payloads that exploit it. To retrieve data, you can use UNION-based SQL injection, which allows you to combine your query with legitimate database queries executed by the application. This technique helps extract data from other tables and present it in the application's response. Time-based blind SQL injection is another method that leverages the delay in the application's response to infer the validity of injected queries. By injecting sleep statements or conditional statements, you can gather information about the database structure and contents. Error-based SQL injection exploits error messages generated by the database management system. By crafting payloads that trigger specific errors, you can extract valuable information, such as database version, table names, or even sensitive data. Blind SQL injection attacks are particularly challenging because they don't reveal information directly in the application's response. To succeed, you must craft payloads that infer data* through Boolean-based or time-based techniques. Boolean-based SQL injection involves injecting queries that return either true or false results. By observing changes in the application's behavior or response times, you can deduce information about the database. Time-based blind SQL injection, as mentioned earlier, relies on time delays to determine the validity of injected queries. You can exploit this vulnerability to extract data progressively. Once you've successfully exploited an SQL injection vulnerability and retrieved data from the database, you can escalate your attack by manipulating the database contents. For instance, you can perform data manipulation operations, such as inserting, updating, or deleting records. These actions can have severe consequences, especially if you modify critical data or create new accounts with administrative privileges. To further hone your SQL injection skills, it's crucial to stay updated on database management system-specific quirks and vulnerabilities. Different database systems, like MySQL, PostgreSQL, Oracle, or Microsoft SQL Server, may exhibit unique behavior when subjected to SQL injection attacks. Knowing these nuances can help you craft more effective payloads and increase your chances of success. Additionally, you should familiarize yourself with evasion techniques and defensive mechanisms employed by modern web applications. Web application firewalls (WAFs) and input validation filters aim to detect and block SQL injection attempts. To bypass these defenses, you can use encoding, obfuscation, or evasion techniques to disguise your payloads. For example, you can encode your malicious payloads in different character sets, such as URL encoding, hexadecimal encoding, or double URL encoding. This can confuse WAFs and make your payloads more challenging to detect. Using comment characters or line breaks can help break up your payload and bypass input validation filters. You can also leverage time delays to slow down your SQL injection attacks and evade detection. It's important to note that ethical hacking and penetration testing should only be performed on systems and applications for which you have explicit authorization. Unauthorized penetration testing is illegal and unethical. Before attempting any SQL injection attacks, always obtain proper authorization and ensure you are operating within the boundaries of the law. To summarize, exploiting SQL injection vulnerabilities requires a deep understanding of SQL databases, crafting effective payloads, and bypassing defensive mechanisms. Mastering this skill is crucial for penetration testers and security professionals to uncover vulnerabilities in web applications and help organizations secure their systems effectively. In the next chapter, we will explore another common web application vulnerability: Cross-Site Scripting (XSS) attacks and how to exploit them effectively.

Chapter 6: Client-Side Attacks and Beyond

Next, we will delve into the art of crafting and executing client-side attacks, a fundamental skill for penetration testers and security professionals. Client-side attacks target the vulnerabilities in the software running on a user's device, such as web browsers, email clients, or office applications. These attacks can be highly effective for compromising the security of a target, as they often rely on the end user's actions or lack of security awareness. To successfully craft and execute client-side attacks, you need to understand the various attack vectors, vulnerability types, and the techniques required for exploitation. One of the most common types of client-side attacks is known as "Phishing." Phishing attacks involve tricking users into disclosing sensitive information, such as login credentials, credit card numbers, or personal information, by posing as a trustworthy entity. These attacks are usually delivered through email, instant messages, or deceptive websites that closely mimic legitimate ones. As a penetration tester or security professional, you can simulate phishing attacks to assess an organization's vulnerability to social engineering. To craft convincing phishing emails, you must understand the psychology of social engineering and create messages that lure recipients into taking specific actions. Attackers often employ urgency, curiosity, or fear tactics to manipulate users into clicking malicious links or opening malicious *attachments. It's crucial to create visually convincing emails and websites that mimic the brand and appearance of legitimate organizations. Using email templates or web pages that closely resemble those of well-known companies can increase the effectiveness of your phishing simulations. Another common client-side attack vector is "Drive-By Downloads," which involves exploiting vulnerabilities in a user's web browser or plugins without their knowledge or consent. Attackers often host malicious code on compromised or malicious websites, and when users visit these sites, their browsers are attacked. Exploiting browser vulnerabilities allows attackers to download and execute malicious payloads on the victim's system. As a penetration tester, you can craft malicious web pages and payloads to assess an organization's vulnerability to drive-by download attacks. To effectively execute drive-by download attacks, you need to identify and exploit browser vulnerabilities or plugin vulnerabilities, such as those found in Adobe Flash or Java. Exploiting these vulnerabilities requires a deep understanding of web technologies and how browsers handle various content types. Client-side attacks can also involve exploiting vulnerabilities in document formats, such as PDFs or Microsoft Office documents. Attackers may embed malicious code or scripts within these documents to compromise the user's system when the document is opened. Penetration testers should be well-versed in crafting malicious documents and understanding how to exploit vulnerabilities in popular document readers. Understanding the intricacies of file format* vulnerabilities and how to weaponize them is essential for conducting successful client-side attacks. Social Engineering is a critical aspect of client-side attacks, as attackers often rely on manipulating users' emotions and behaviors to achieve their goals. As a penetration tester, you should be proficient in social engineering techniques to assess an organization's susceptibility to these types of attacks. Engaging in pretexting, baiting, tailgating, or other social engineering tactics can help uncover vulnerabilities in an organization's human factor security defenses. Client-side attacks can also leverage malicious scripts or code embedded within legitimate websites, a technique known as "Watering Hole Attacks." In these attacks, attackers compromise a website frequently visited by the target audience, making it likely that victims will be exposed to the malicious code. This approach can be particularly effective against specific industries or groups of users. As a penetration tester, simulating watering hole attacks requires identifying high-value target websites, understanding their technologies, and crafting malicious payloads. You must also be proficient in identifying and exploiting web vulnerabilities, such as Cross-Site Scripting (XSS) or Remote File Inclusion (RFI), to deliver your payloads. Another client-side attack vector is "Malvertising," which involves delivering malicious code through online advertisements. Attackers may compromise ad networks to distribute malicious ads that can lead users to malicious websites or exploit vulnerabilities in their browsers or plugins. To simulate malvertising attacks, you need to understand how to *create and distribute malicious ads, as well as exploit browser or plugin vulnerabilities when users view or click on these ads. Additionally, you should be familiar with web tracking and advertising technologies to identify potential targets for malvertising campaigns. The success of client-side attacks often depends on the attacker's ability to evade detection by security mechanisms, such as antivirus software or intrusion detection systems. To do this, attackers may use various obfuscation techniques to hide their malicious payloads from security scanners. As a penetration tester, you can employ these same techniques to evaluate an organization's security posture. Understanding code obfuscation, encoding, or encryption methods used by attackers can help you assess the effectiveness of an organization's defenses against client-side attacks. In summary, crafting and executing client-side attacks require a deep understanding of various attack vectors, vulnerability types, and social engineering tactics. As a penetration tester or security professional, mastering these skills is essential for assessing an organization's susceptibility to client-side threats. In the next chapter, we will explore another critical aspect of web application security: Cross-Site Scripting (XSS) attacks and how to defend against them effectively. Next, we will venture into the fascinating realm of client-side security, an essential component of modern cybersecurity that extends beyond web browsers. Client-side security encompasses a wide array of considerations related to the security of software and applications running on a user's device. While web browsers are a significant focus, client-side security also includes the security of email clients, mobile apps, desktop applications, and even Internet of Things (IoT) devices. Understanding client-side security is crucial for both penetration testers and security professionals to identify and mitigate vulnerabilities effectively. One of the central pillars of client-side security is securing the software that users interact with daily. This includes web browsers like Chrome, Firefox, Edge, Safari, and others, as well as the plugins and extensions that extend their functionality. Browsers are a prime target for attackers due to their widespread use and the potential for exploiting vulnerabilities to compromise a user's system. As a penetration tester, it's essential to understand browser security settings, configuration best practices, and common vulnerabilities that could lead to client-side attacks. By assessing browser security, you can help organizations defend against attacks such as drive-by downloads, malicious extensions, and cross-site scripting (XSS) attacks. Email clients are another critical component of client-side security, as email remains a primary vector for delivering malware and phishing attacks. Securing email clients involves configuring settings to block potentially malicious content, scanning attachments for malware, and implementing strong authentication mechanisms. Penetration testers should assess email client configurations and test their resistance to phishing and email-based attacks to identify areas for improvement. Mobile applications have become an integral part of our daily lives, and their security is paramount, especially as mobile devices often* contain sensitive personal and corporate information. Assessing the security of mobile apps requires a deep understanding of mobile operating systems (iOS and Android), app permissions, data encryption, and secure coding practices. Mobile app penetration testing involves identifying vulnerabilities like insecure data storage, broken authentication, or code injection, which could lead to data breaches or unauthorized access. In addition to web browsers, email clients, and mobile apps, client-side security extends to desktop applications used for various purposes. These applications include word processors, spreadsheets, image editors, and more. Penetration testers must evaluate the security of these applications to ensure they don't expose users to risks such as macro-based attacks, malicious documents, or vulnerable plugins. Client-side security also encompasses the growing ecosystem of Internet of Things (IoT) devices, which range from smart thermostats and doorbells to industrial sensors and medical devices. Securing IoT devices is challenging due to their diverse nature and limited resources, but it's vital to assess their security, as vulnerabilities can have severe consequences. Penetration testers need to understand the unique challenges posed by IoT security, such as weak default credentials, lack of update mechanisms, and potential attack vectors through unsecured communication channels. While we've covered various aspects of client-side security, it's essential to recognize that user awareness and education play a significant role in overall security. Even the most secure software and devices can be compromised if users fall victim to social engineering attacks like phishing or downloading malicious attachments. Penetration testers often simulate these attacks to assess an organization's vulnerability to social engineering and user manipulation. By crafting convincing phishing emails, testing user responses, and conducting security awareness training, testers can help organizations bolster their client-side security defenses. Another crucial aspect of client-side security is the timely application of security patches and updates. Software vendors regularly release patches to address known vulnerabilities, and failing to apply these updates promptly can leave systems exposed to attacks. As a penetration tester, it's important to assess an organization's patch management practices to identify potential weaknesses and vulnerabilities. Beyond evaluating and testing client-side security, penetration testers and security professionals must stay informed about emerging threats and attack techniques. The world of cybersecurity is ever-evolving, with attackers continuously devising new methods to exploit client-side vulnerabilities. To remain effective in protecting systems and data, security experts must keep their knowledge up-to-date through continuous learning, industry research, and collaboration with peers. In summary, client-side security is a multifaceted domain that encompasses various software and devices used by individuals and organizations. Understanding and effectively securing client-side components, such as web browsers, email clients, mobile apps, desktop applications, and IoT devices, is essential for *safeguarding against a wide range of threats. Penetration testers and security professionals play a vital role in assessing and improving client-side security by identifying vulnerabilities, conducting simulations, and providing guidance on best practices. By staying vigilant, informed, and proactive, we can collectively enhance client-side security and protect against evolving cyber threats.*

Chapter 7: Network and Infrastructure Hacking with Burp Suite

Next, we delve into the essential process of assessing network and infrastructure weaknesses, a critical component of any comprehensive penetration testing effort. The network and infrastructure of an organization serve as the backbone for all its digital operations and data storage, making them attractive targets for attackers. To effectively evaluate these weaknesses, penetration testers must adopt a structured approach that includes a thorough understanding of the target environment. Network assessment typically begins with reconnaissance, where testers gather information about the organization's network topology, IP address ranges, and the systems in use. This phase is critical as it helps testers identify potential entry points and attack surfaces within the network. Reconnaissance techniques range from passive approaches like DNS enumeration and WHOIS queries to more active methods like network scanning and port enumeration. Once the initial reconnaissance is complete, penetration testers move on to vulnerability scanning and assessment. This step involves using tools and techniques to identify known vulnerabilities and weaknesses in the network and infrastructure components. Common tools like Nessus, OpenVAS, and Qualys can assist testers in discovering vulnerabilities in systems, services, and applications. While automated scanning tools are valuable, manual testing is also crucial, as it allows testers to uncover complex vulnerabilities that automated scanners may miss. After identifying vulnerabilities, the next step is exploitation testing, where penetration testers attempt to exploit the identified weaknesses to gain unauthorized access or escalate privileges within the network. Successful exploitation provides critical insights into the impact of vulnerabilities and the potential risks they pose to the organization. Exploitation techniques can vary widely, from straightforward attacks like brute force password cracking to more advanced methods like buffer overflow exploits. Throughout the testing process, it is essential to maintain strict ethical boundaries and adhere to a defined scope of work to avoid causing any harm to the target organization's systems or data. One of the primary goals of network and infrastructure assessment is to assess the resilience of security controls and countermeasures in place. This includes evaluating the effectiveness of firewalls, intrusion detection systems (IDS), intrusion prevention systems (IPS), and access control lists (ACLs) in mitigating threats. Penetration testers simulate real-world attack scenarios to determine whether these security measures can withstand various intrusion attempts. For example, they might attempt to bypass a firewall rule to gain access to a protected network segment or evade detection by an IDS. Another essential aspect of network assessment is the review of network configurations and policies. This includes analyzing router and switch configurations, firewall rules, and network segmentation to ensure they *align with security best practices. Misconfigured devices or overly permissive rules can create security weaknesses that attackers could exploit. By identifying and addressing these issues, organizations can significantly improve their network security posture. In addition to testing the security of networked systems, penetration testers must also assess the physical security of infrastructure components, such as data centers, server rooms, and networking equipment. Physical security assessments involve evaluating measures like access control, surveillance, and environmental controls. Testers may attempt to gain unauthorized physical access to secure areas or assess the resilience of security mechanisms like biometric locks and surveillance cameras. In some cases, vulnerabilities discovered during physical security assessments can have severe implications for overall network security. Wireless networks are another crucial aspect of network and infrastructure assessment. With the proliferation of Wi-Fi-enabled devices, securing wireless networks is paramount. Penetration testers evaluate the security of Wi-Fi networks by attempting to exploit weaknesses in encryption protocols, authentication mechanisms, and access control. This includes testing for vulnerabilities like weak WEP or WPA passwords, rogue access points, and inadequate encryption. Additionally, testers may assess the effectiveness of network segmentation by attempting to pivot from a compromised wireless network to the wired infrastructure. Ultimately, the goal is to identify vulnerabilities in the wireless environment that could allow unauthorized access to the network.*

Network and infrastructure assessment also extend to cloud environments, which have become an integral part of modern IT infrastructure. Organizations increasingly rely on cloud services like Amazon Web Services (AWS), Microsoft Azure, and Google Cloud Platform (GCP) for scalability and flexibility. However, securing cloud resources presents its own set of challenges. Penetration testers assess cloud security by examining configurations, permissions, and access controls within cloud environments. Common cloud security issues include misconfigured access policies, exposed storage buckets, and inadequate identity and access management (IAM) controls. By identifying these vulnerabilities, testers help organizations protect sensitive data and maintain a strong security posture in the cloud. Throughout the entire network and infrastructure assessment process, communication and collaboration with the organization's IT and security teams are essential. Testers should work closely with these teams to validate findings, prioritize vulnerabilities, and provide guidance on remediation. A transparent and cooperative approach ensures that the assessment aligns with the organization's objectives and facilitates the implementation of effective security measures. In summary, assessing network and infrastructure weaknesses is a fundamental aspect of penetration testing, helping organizations identify and mitigate vulnerabilities that could be exploited by malicious actors. By conducting thorough reconnaissance, vulnerability scanning, exploitation testing, and physical security assessments, penetration *testers play a crucial role in enhancing an organization's security posture. The continuous evolution of technology and emerging threats makes network and infrastructure assessment an ongoing process, requiring organizations to remain vigilant and proactive in addressing security weaknesses.*

Chapter 8: Beyond Web: Cloud, Mobile, and IoT Security

Cloud computing has transformed the way organizations manage and deliver IT services, offering scalability, flexibility, and cost-effectiveness. However, this shift to the cloud has also introduced a new set of security challenges that organizations must address to protect their data and applications. One of the primary concerns in cloud environments is data security, as organizations entrust cloud providers with sensitive information. Data breaches can have severe consequences, including financial losses, regulatory fines, and damage to an organization's reputation. To mitigate this risk, organizations must implement robust encryption mechanisms, both for data in transit and at rest. Another significant challenge is identity and access management (IAM) in the cloud. Ensuring that only authorized individuals and systems can access cloud resources is essential. Inadequate IAM practices can lead to unauthorized access, data leaks, and account compromises. Organizations should implement strong authentication mechanisms, least privilege access policies, and regular access reviews to address these concerns. Securing cloud infrastructure is another critical aspect of cloud security. Misconfigured cloud services and resources can expose organizations to vulnerabilities and attacks. Cloud providers offer a wide range of security features, but it's the responsibility of the organization to configure them correctly. Penetration testing and security assessments can help identify misconfigurations and vulnerabilities in cloud environments. One emerging security challenge in cloud environments is the shared responsibility model. In the shared responsibility model, cloud providers are responsible for the security of the cloud infrastructure, while customers are responsible for securing their data and applications within that infrastructure. Understanding and properly defining these responsibilities is essential to avoid security gaps. Next, organizations must address the challenge of visibility and monitoring. Traditional on-premises security tools may not provide the same level of visibility in the cloud. Organizations need to invest in cloud-native security solutions that can monitor and analyze cloud traffic, events, and logs. This visibility is crucial for detecting and responding to security incidents. Cloud environments also introduce challenges related to compliance and regulatory requirements. Different regions and industries have specific data protection and privacy regulations that organizations must adhere to. Ensuring compliance in the cloud requires a combination of proper configurations, data classification, and auditing capabilities. One of the most significant security challenges in cloud environments is the rapid pace of change. Cloud services and features are constantly evolving, which means that security teams must adapt quickly to new threats and vulnerabilities. Security practices that worked yesterday may not be effective tomorrow. Organizations need to invest in *ongoing security training and awareness programs for their teams. Another challenge in cloud security is the management of keys and secrets. Cloud applications often rely on cryptographic keys and secrets to encrypt data and authenticate users. Organizations must have robust key management practices in place to protect these critical assets. Additionally, organizations should implement multi-factor authentication (MFA) for accessing cloud resources. MFA adds an extra layer of security by requiring users to provide multiple forms of authentication before gaining access. Social engineering attacks are another concern in cloud environments. Attackers may attempt to trick users or employees into revealing sensitive information or credentials. Security awareness training is essential to educate users about the risks of social engineering and how to recognize and respond to such attacks. As organizations adopt a multi-cloud or hybrid cloud approach, they face the challenge of ensuring consistent security policies and controls across different cloud providers and on-premises environments. This requires careful planning, the use of cloud security best practices, and the adoption of security automation tools. Finally, organizations must prepare for the possibility of a cloud-specific incident response plan. While traditional incident response plans are valuable, they may not cover all aspects of cloud-related incidents. A dedicated cloud incident response plan can help organizations respond effectively to cloud security breaches. In summary, the transition to cloud computing has brought about significant benefits for organizations, but it has also introduced a unique set of* security challenges. From data security and IAM to compliance and the shared responsibility model, addressing these challenges requires a proactive and holistic approach to cloud security. By staying informed, implementing best practices, and continually adapting to the evolving threat landscape, organizations can enhance their security posture in cloud environments. Mobile and Internet of Things (IoT) devices have become an integral part of our daily lives, offering convenience and connectivity like never before. From smartphones and tablets to smart thermostats and wearable fitness trackers, these devices have transformed the way we interact with technology. However, with this increased connectivity comes an expanded attack surface, making it crucial to assess the security of mobile and IoT devices. One powerful tool for assessing the security of these devices is Burp Suite, a versatile and comprehensive penetration testing toolkit. Next, we will explore how Burp Suite can be used to assess the security of mobile and IoT devices, helping you identify vulnerabilities and enhance their security. Mobile devices, such as smartphones and tablets, are ubiquitous in today's world, and they store a vast amount of personal and sensitive information. Ensuring the security of these devices is essential to protect users' privacy and prevent data breaches. Burp Suite can be employed to assess the security of mobile applications installed on these devices. Mobile application security assessments involve testing the applications for vulnerabilities that could be exploited by malicious actors. One common approach is to perform dynamic analysis, where Burp Suite intercepts and analyzes the traffic between the mobile application and its backend server. This allows you to identify potential security flaws, such as insecure communication, improper session management, or even vulnerabilities in the application's code. Another valuable feature of Burp Suite is its ability to analyze the APIs used by mobile applications. APIs (Application Programming Interfaces) are the mechanisms that allow mobile applications to interact with backend servers and services. By intercepting and inspecting API calls made by the mobile app, you can uncover vulnerabilities or weaknesses that might be exploited by attackers. For example, you might discover that an API endpoint does not require proper authentication or that it lacks proper input validation, leaving it susceptible to injection attacks. IoT devices, on the other hand, have gained popularity for their ability to automate tasks and provide real-time data, from smart thermostats that regulate home temperatures to connected cameras that enhance security. However, the proliferation of IoT devices has also introduced security risks, as many of these devices have limited security features and are often connected to the internet. Burp Suite can be used to assess the security of IoT devices by testing the communication protocols and APIs they use to interact with other devices and services. One common scenario is assessing the security of an IoT device's companion mobile application, which often serves as the interface for users to control and monitor the device. Using Burp Suite, you can analyze the communication between the mobile app and the IoT device, looking for vulnerabilities *or weaknesses that could be exploited. For example, you might discover that the IoT device communicates with the mobile app over an insecure channel, potentially exposing sensitive data to eavesdroppers. Additionally, Burp Suite can be employed to assess the security of the APIs that IoT devices use to communicate with cloud services or other devices. By intercepting and analyzing these API requests and responses, you can identify security flaws, such as missing authentication, weak encryption, or inadequate input validation. One of the challenges in assessing mobile and IoT device security is the diversity of platforms, operating systems, and communication protocols involved. Different mobile devices run on various operating systems, including iOS and Android, each with its unique security features and challenges. IoT devices often use a variety of communication protocols, such as Wi-Fi, Bluetooth, Zigbee, or cellular networks, making it essential to adapt your testing approach accordingly. Burp Suite offers flexibility in this regard, as it supports various proxy configurations and settings to intercept and analyze traffic from a wide range of devices and applications. To assess the security of mobile applications on different platforms, you may need to configure your mobile device to route its traffic through Burp Suite's proxy. This setup allows Burp Suite to intercept and inspect the communication between the mobile app and its backend services. For iOS devices, you can configure the device to trust Burp Suite's SSL certificate, enabling you to intercept HTTPS traffic as well. On Android devices, you can use the Burp Suite mobile app or manually configure the device to use Burp Suite as a proxy. For IoT devices, the testing approach may vary depending on the device's communication methods. For devices that use Wi-Fi or Ethernet for connectivity, you can configure your Burp Suite instance to intercept and analyze network traffic between the device and its backend services. If the IoT device uses Bluetooth or Zigbee, you may need specialized tools and hardware to capture and analyze the communication. Regardless of the specific setup, Burp Suite's flexibility and proxy capabilities make it a valuable tool for assessing the security of mobile and IoT devices. In addition to dynamic analysis, Burp Suite can also perform static analysis of mobile applications. Static analysis involves examining the application's code and binaries without executing them. This can help identify vulnerabilities, such as hardcoded credentials, insecure storage of sensitive data, or improper handling of user inputs. Burp Suite integrates with various static analysis tools, allowing you to automate this process and generate comprehensive reports. Furthermore, Burp Suite offers automation features that can streamline the testing process for mobile and IoT devices. You can create custom scripts and macros to automate repetitive tasks, such as login sequences or data submission. This automation can significantly speed up the testing process, especially when assessing multiple mobile applications or IoT devices. Additionally, Burp Suite's extensibility allows you to create custom plugins and extensions tailored to your specific testing needs. For example, you can develop a custom extension to test the security of an IoT device's proprietary communication protocol. These extensions enhance the capabilities of Burp Suite and enable you to address unique challenges in mobile and IoT device security assessments. When assessing mobile and IoT devices with Burp Suite, it's essential to consider the potential impact on users and the device's functionality. Testing should be conducted in a controlled environment, and you should obtain proper authorization from the device owner or application developer. Furthermore, thorough documentation of the testing process, findings, and recommendations is crucial for communicating the results to stakeholders and developers. In summary, mobile and IoT devices have revolutionized the way we interact with technology, but they also introduce new security challenges. Assessing the security of these devices is essential to protect user data and ensure the integrity of IoT ecosystems. Burp Suite's versatile features, including dynamic and static analysis, automation, and extensibility, make it a valuable tool for assessing the security of mobile and IoT devices. By using Burp Suite effectively and adapting your testing approach to different platforms and communication methods, you can identify vulnerabilities and enhance the security of these devices.*

Chapter 9: Burp Suite in Enterprise Environments

Scaling Burp Suite for enterprise use is a critical consideration for organizations looking to enhance their application security. As businesses grow, they typically deploy more web applications, APIs, and services, which increases their attack surface and vulnerability. Burp Suite, with its extensive set of features, can be a valuable asset in securing these assets, but it's crucial to scale it effectively to meet the demands of the enterprise. One of the primary challenges when scaling Burp Suite is managing the increasing number of web applications and APIs that require security testing. In a growing enterprise, it's common to have dozens or even hundreds of web applications and services, each with its unique requirements and testing schedules. To address this challenge, organizations can implement a centralized management system for Burp Suite, which allows security teams to coordinate, schedule, and monitor scans across multiple applications. This centralized approach streamlines the management of scanning targets, ensures consistent testing methodologies, and facilitates reporting and remediation efforts. Another aspect of scaling Burp Suite is accommodating the higher volume of scanning tasks and the associated resource demands. As the number of applications and services to be tested grows, organizations may need to invest in more powerful hardware to support concurrent scans. This can include high-performance servers with ample processing power, *memory, and storage capacity. Additionally, cloud-based solutions can provide scalability by allowing organizations to spin up additional scanning instances as needed. It's essential to allocate sufficient resources to meet the demands of concurrent scans without compromising on scan quality or speed. Furthermore, organizations can enhance their scalability by leveraging Burp Suite's automation capabilities. By automating routine tasks, such as scan scheduling, result analysis, and reporting, security teams can increase their testing efficiency and focus on critical vulnerabilities and issues. Automation can also help in managing large-scale scanning efforts, especially when dealing with frequent application updates or changes. In addition to handling a higher volume of scanning tasks, enterprises must consider the distributed nature of their environments. Large organizations often have multiple branches, subsidiaries, or business units, each with its web applications and services. Scaling Burp Suite to accommodate these distributed environments requires a well-defined strategy for deployment and collaboration. One approach is to establish a Burp Suite deployment at the organizational level, with multiple scanning nodes distributed across different locations or business units. These scanning nodes can be configured to share scan results and collaborate on security assessments while maintaining a centralized management and reporting structure. Moreover, organizations can tailor their scanning configurations based on the specific needs of each business unit, allowing for customization while still adhering to corporate security standards. Scaling Burp*

Suite also involves handling large volumes of scan data and results. As the number of scans increases, so does the amount of data generated, including scan reports, issues, and remediation recommendations. Effective data management and storage solutions are essential to prevent data overload and ensure easy access to historical scan results. Organizations can implement a structured approach to data storage, including archival policies, categorization of scan data, and integration with security information and event management (SIEM) systems. This enables efficient data retention, retrieval, and analysis, facilitating continuous improvement of security practices. Another critical consideration in scaling Burp Suite for enterprise use is collaboration and communication within security teams. Large organizations often have multiple security professionals, each with their expertise and responsibilities. Efficient collaboration and knowledge sharing are vital to ensure that security assessments are comprehensive and vulnerabilities are addressed promptly. Burp Suite offers features like team collaboration and integration with issue tracking systems, enabling security teams to work together seamlessly. These collaborative tools help in assigning, tracking, and resolving security issues, ensuring that vulnerabilities are addressed efficiently. Additionally, they allow for knowledge sharing and the transfer of expertise among team members. Furthermore, as organizations scale their security testing efforts, they may need to implement processes for integrating Burp Suite with other security tools and systems. This can *include integrating with vulnerability management systems, continuous integration/continuous deployment (CI/CD) pipelines, or orchestration platforms. Integration enables a smoother workflow, where scan results can be automatically fed into remediation processes or trigger alerts for immediate action. It also ensures that security testing is an integral part of the software development lifecycle, promoting a proactive approach to security. Lastly, organizations scaling Burp Suite for enterprise use must prioritize the ongoing training and development of their security teams. As the complexity of security assessments and the number of scanning targets increase, it's essential that security professionals remain up-to-date with the latest security trends, techniques, and best practices. Organizations can invest in training programs, certifications, and knowledge sharing sessions to empower their security teams. This ongoing education ensures that security teams can effectively utilize Burp Suite's advanced features and adapt to evolving security challenges. In summary, scaling Burp Suite for enterprise use is a multifaceted endeavor that involves centralizing management, allocating resources effectively, automating tasks, addressing distributed environments, managing data, fostering collaboration, integrating with other security tools, and investing in team development. By carefully planning and implementing these strategies, organizations can harness the power of Burp Suite to secure their growing web application and API portfolios effectively.*

Chapter 10: Reporting, Remediation, and Staying Ahead

Effective reporting in the context of cybersecurity is a crucial aspect of communicating the results of security assessments and penetration tests to various stakeholders. When conducting security assessments or penetration tests, the primary objective is to identify vulnerabilities, weaknesses, and potential threats within an organization's systems and applications. However, the value of these assessments lies not only in finding vulnerabilities but also in effectively communicating the findings and recommendations to the relevant stakeholders. Stakeholders can include executives, IT teams, developers, compliance officers, and even external auditors, each with different priorities and levels of technical expertise. Tailoring the reporting process to meet the needs of these diverse stakeholders is essential to ensure that the findings are understood, acted upon, and aligned with the organization's overall security objectives. The first step in effective reporting is to understand the specific requirements and expectations of each stakeholder group. For example, executives and senior management may be more concerned with high-level summaries and strategic implications, while technical teams may require detailed technical information to remediate vulnerabilities. By identifying these needs early in the process, you can customize your reporting approach accordingly. One common approach is to provide multiple levels of reporting, each tailored to a different audience. For executives and senior management, a high-level executive summary can be a valuable tool. This summary should provide a concise overview of the assessment's objectives, key findings, potential business impacts, and high-level recommendations. The executive summary should avoid technical jargon and focus on the business implications of the findings. Additionally, it can include an overall risk assessment that helps executives prioritize remediation efforts and allocate resources effectively. Technical teams, on the other hand, require more detailed and technical information. For these stakeholders, the full assessment report should provide in-depth descriptions of vulnerabilities, including the technical details of how they were discovered, their potential impact, and specific remediation recommendations. Including proof-of-concept code, screenshots, and other technical evidence can assist technical teams in understanding and addressing the issues effectively. While detailed information is vital for technical teams, it's equally important to provide actionable recommendations that guide remediation efforts. Rather than just highlighting problems, the report should offer clear and practical guidance on how to fix identified vulnerabilities. These recommendations should be prioritized based on risk, potential impact, and ease of remediation. By providing a clear roadmap for addressing vulnerabilities, technical teams can streamline their efforts and make meaningful improvements to the organization's security posture. In addition to the executive summary and detailed technical report, it's often helpful to include an appendix or supplementary materials. These can include raw data, logs, scan results, and additional technical details that may be of interest to specific stakeholders. Including supplementary materials allows stakeholders to dig deeper into the findings if needed, ensuring transparency and thoroughness in the reporting process. Furthermore, timelines and deadlines are crucial aspects of effective reporting. Stakeholders need to know when assessments were conducted, when vulnerabilities were discovered, and when remediation efforts should be completed. Clearly defining timelines and deadlines helps set expectations and ensures that identified issues are addressed promptly. In some cases, compliance requirements or regulatory standards may dictate specific reporting timelines, so it's essential to adhere to these guidelines. Additionally, reporting should not be a one-time event. Regular, ongoing reporting is essential to track progress, measure improvements, and ensure that security concerns are addressed systematically. Regularly scheduled security assessments and penetration tests, followed by consistent reporting, help organizations maintain a proactive security posture and reduce the risk of security incidents. Moreover, effective reporting should not be limited to written documents. Visual aids, such as charts, graphs, and dashboards, can enhance the communication of complex information. Visual representations of vulnerability trends, risk assessments, and remediation progress can make it easier for stakeholders to grasp the overall security status at a glance. For example, a dashboard could show the number of high-risk vulnerabilities over time, illustrating whether the organization's security posture is improving or deteriorating. Visualization tools can help executives and technical teams alike make data-driven decisions and prioritize security efforts effectively. Another aspect to consider is the language used in reporting. It's crucial to strike a balance between technical accuracy and accessibility. While technical teams may appreciate detailed technical language, executives and non-technical stakeholders may struggle to understand it. Using plain language and avoiding excessive technical jargon can ensure that the report is accessible to a broader audience. However, it's essential not to oversimplify or omit critical technical details that could impact decision-making or remediation efforts. In some cases, it may be necessary to provide both simplified and technical versions of the report to cater to different stakeholders' needs. Lastly, reporting should not be limited to pointing out problems. Acknowledging successes and improvements in the security posture is equally important. By highlighting areas where the organization has made progress or successfully mitigated vulnerabilities, you can motivate teams and demonstrate the value of security efforts. This positive reinforcement can encourage ongoing dedication to security and promote a culture of continuous improvement. In summary, effective reporting for different stakeholders involves understanding their unique requirements, tailoring the reporting approach to meet those needs, providing clear timelines and deadlines, using visual aids, striking the right balance between technical and plain language, and recognizing and celebrating successes. By adopting a comprehensive and customized reporting approach, organizations can ensure that the results of security assessments and penetration tests are communicated effectively and result in meaningful actions to improve security posture. In the dynamic and ever-evolving field of cybersecurity, one thing is certain: learning is not a one-time event, but a continuous journey. Cyber threats are constantly changing and becoming more sophisticated, making it imperative for security professionals to stay ahead of the curve. The ability to adapt, learn, and evolve is not just a skill but a mindset that can significantly impact an organization's security posture.

Continuous learning in cybersecurity encompasses a broad spectrum of activities, from keeping up with the latest threat trends to acquiring new technical skills and understanding emerging technologies. Let's explore why continuous learning is essential and how to effectively stay ahead in the realm of security.

Understanding the Need for Continuous Learning

Cybersecurity threats are relentless, with attackers continually developing new tactics and exploiting vulnerabilities. The moment a security professional becomes complacent or stops learning, they risk falling behind in an ever-advancing battlefield.

One of the primary reasons for continuous learning is the evolving nature of technology. As organizations adopt new technologies and platforms, they introduce new attack vectors and potential vulnerabilities. Staying informed about these technological advancements is crucial for identifying and mitigating security risks effectively.

Moreover, the threat landscape is in a constant state of flux. New attack techniques, malware strains, and vulnerabilities are discovered regularly. Security professionals need to stay informed about the latest threat intelligence to defend against these emerging risks effectively.

How to Stay Ahead in Security Through Continuous Learning

Engage in Ongoing Training and Certification: Enroll in relevant courses, certifications, and training programs to acquire new skills and knowledge. Certifications like Certified Information Systems Security Professional (CISSP), Certified Ethical Hacker (CEH), and Certified Information Security Manager (CISM) are highly regarded in the cybersecurity field.

Attend Conferences and Workshops: Participate in industry conferences, workshops, and seminars. These events provide opportunities to network with peers, learn from experts, and gain insights into the latest security trends and technologies.

Read Books, Blogs, and Research Papers: Stay informed by reading books, blogs, and research papers authored by security experts. Books such as "The Web Application Hacker's Handbook" and "Hacking: The Art of Exploitation" offer in-depth knowledge on various aspects of cybersecurity.

Follow Security News and Blogs: Regularly follow reputable security news websites and blogs that report on the latest cyber threats, breaches, and vulnerabilities. Staying updated on current events in the cybersecurity world is crucial.

Participate in Capture The Flag (CTF) Competitions: CTF competitions provide hands-on experience in solving security challenges and puzzles. They are an excellent way to hone your technical skills and gain practical knowledge.

Join Security Communities and Forums: Engage with online security communities and forums where professionals discuss security issues, share insights, and seek help. Platforms like Reddit's r/netsec and Stack Exchange's Security Stack Exchange are valuable resources.

Experiment with Home Labs: Set up a home lab environment where you can experiment with different security tools, techniques, and technologies. Hands-on experience is invaluable for learning and mastering cybersecurity skills.

Mentorship and Peer Learning: Seek mentorship from experienced security professionals and collaborate with peers. Learning from others' experiences and insights can accelerate your growth.

Stay Informed About Regulations and Compliance: Understand the regulatory landscape in your industry and the compliance requirements that organizations must meet. Compliance is an integral part of cybersecurity, and staying updated on relevant regulations is essential.

Ethical Hacking and Vulnerability Research: Consider ethical hacking and vulnerability research as a way to learn more about real-world security issues. Organizations often hire ethical hackers to identify and mitigate vulnerabilities in their systems.

Stay Adaptable and Curious: Cultivate a growth mindset by remaining adaptable and curious. Embrace challenges and view failures as opportunities for learning and improvement.

Build a Professional Network: Establish and maintain a strong professional network in the cybersecurity community. Networking can open doors to new opportunities and provide valuable insights.

The Role of Soft Skills in Continuous Learning

While technical skills are critical in cybersecurity, soft skills also play a vital role in professional growth. Effective communication, problem-solving, critical thinking, and teamwork are essential for security professionals to navigate complex challenges successfully.

Cybersecurity incidents often require coordination across various teams within an organization, and the ability to communicate technical information clearly to non-technical stakeholders is invaluable. Moreover, ethical dilemmas and decision-making in the field of security can benefit from strong critical thinking and problem-solving skills.

Conclusion: A Lifelong Learning Journey

In the realm of cybersecurity, continuous learning is not merely a career choice; it's a necessity. As technology evolves and cyber threats become more sophisticated, staying ahead in security requires a commitment to ongoing education and adaptation. Embracing this *journey of lifelong learning empowers security professionals to protect organizations effectively and contribute to a safer digital world. So, whether you're just starting your cybersecurity career or have years of experience, remember that the path to excellence in security is paved with a dedication to continuous learning and improvement.*

Conclusion

In this comprehensive book bundle, "Burp Suite: Novice to Ninja - Pen Testing Cloud, Network, Mobile & Web Applications," we embarked on an exciting journey through the fascinating world of web application security and beyond. Across four distinct volumes, we delved deep into the realm of ethical hacking, penetration testing, and security auditing with Burp Suite as our trusty companion.

In "Book 1 - Burp Suite Fundamentals: A Novice's Guide to Web Application Security," we laid the foundation for our exploration. We started as novices, gradually gaining a profound understanding of web application vulnerabilities and the essential tools that Burp Suite provides for identifying and mitigating these threats. This book served as our stepping stone into the captivating universe of web application security.

"Book 2 - Mastering Burp Suite: Pen Testing Techniques for Web Applications" propelled us further into the art of ethical hacking. Armed with advanced techniques and in-depth knowledge, we honed our skills in using Burp Suite to uncover vulnerabilities, execute attacks, and secure web applications. We became proficient in both manual and automated testing, making us formidable web security practitioners.

"Book 3 - Penetration Testing Beyond Web: Network, Mobile & Cloud with Burp Suite" expanded our horizons beyond web applications. We ventured into the realms of network, mobile, and cloud security, discovering how Burp Suite could be adapted to address a wider range of challenges. Armed with our newfound expertise, we were ready to tackle diverse security assessments in today's complex technological landscape.

Finally, "Book 4 - Burp Suite Ninja: Advanced Strategies for Ethical Hacking and Security Auditing" elevated us to the status of security auditors and ethical hacking ninjas. We explored advanced strategies, customization, scripting, and automation with Burp Suite. With this knowledge, we not only identified vulnerabilities but also crafted comprehensive security reports and devised effective remediation strategies.

As we conclude this book bundle, we've evolved from novices to security experts, capable of safeguarding web applications, networks, mobile devices, and cloud environments. We've embraced the mindset of ethical hackers, ready to face the evolving landscape of cybersecurity challenges with confidence.

Remember, the journey of learning and mastering Burp Suite and penetration testing is ongoing. The world of cybersecurity is dynamic, with new threats and technologies emerging continuously. Therefore, I encourage you to continue your pursuit of knowledge and expertise, to stay curious, and to remain committed to the principles of ethical hacking and responsible disclosure.

Thank you for joining me on this remarkable journey through "Burp Suite: Novice to Ninja - Pen Testing Cloud, Network, Mobile & Web Applications." May your newfound skills and insights empower you to make a meaningful impact on the security of the digital world.

Printed in the USA
CPSIA information can be obtained
at www.ICGtesting.com
CBHW072118020824
12436CB00146B/1225